Solar Radiation and Daylight Models for the Energy Efficient Design of Buildings
(with software compact disk)

T. Muneer
Napier University

with a chapter on Solar Spectral Radiation
by H. Kambezidis, National Observatory of Athens

Architectural Press

Architectural Press
An imprint of Butterworth-Heinemann
Linacre House, Jordan Hill, Oxford OX2 8DP
A division of Reed Educational and Professional Publishing Ltd

A member of the Reed Elsevier plc group

OXFORD BOSTON JOHANNESBURG
MELBOURNE NEW DELHI SINGAPORE

First published 1997

© Reed Educational and Professional Publishing Ltd 1997

All rights reserved. No part of this publication
may be reproduced in any material form (including
photocopying or storing in any medium by electronic
means and whether or not transiently or incidentally
to some other use of this publication) without the
written permission of the copyright holder except
in accordance with the provisions of the Copyright,
Designs and Patents Act 1988 or under the terms of a
licence issued by the Copyright Licensing Agency Ltd,
90 Tottenham Court Road, London, England W1P 9HE.
Applications for the copyright holder's written permission
to reproduce any part of this publication should be addressed
to the publishers

British Library Cataloguing in Publication Data
A catalogue record for this book is available from the British Library

ISBN 0 7506 2495 7

Library of Congress Cataloguing in Publication Data
A catalogue record for this book is available from the Library of Congress

Printed and bound in Great Britain by
Biddles Ltd, Guildford and King's Lynn

For my parents

        Low in the earth
I lived in the realms of ore and stone;
and then I smiled in many-tinted flowers;
Then roving with the wild and wandering hours,
Over earth and air and ocean's zone,
        In a new birth,
          I dived and flew,
            And crept and ran,
And all the secret of my essence drew
Within a form that brought them all to view –
And then my goal,
Beyond the clouds, beyond the sky,
In angel form; and then away
Beyond the bounds of night and day.

From *Masnavi-ye-Manavi (Spiritual Couplets)*
by Jalaluddin Rumi (1207–73), Persian mystical poet. It is possible to interpret this poem as a eulogy of the sun which represents a 'whirling dervish'.
The sect of the whirling dervishes was founded by Rumi's followers.

CONTENTS

FOREWORD *Professor Peter Tregenza, University of Sheffield* — vii

PREFACE — ix

ACKNOWLEDGEMENTS — xi

LIST OF ELECTRONIC FILES ON COMPACT DISK — xiii

LIST OF FIGURES — xv

LIST OF TABLES — xix

INTRODUCTION — xxi

1	FUNDAMENTALS	1
1.1	Solar day	2
1.2	Equation of time	3
1.3	Apparent solar time	5
1.4	Solar declination	8
1.5	Solar altitude, azimuth and incidence angle	10
1.6	Astronomical sunrise and sunset	13
1.7	Actual sunrise and sunset	13
1.8	Twilight	13
1.9	Distance between two locations	15
1.10	Solar radiation and daylight measurement	16
1.11	Statistical evaluation of models	20
1.12	Exercises	24
	References	24
2	DAILY IRRADIATION	27
2.1	Monthly-averaged daily horizontal global irradiation	28
2.2	Monthly-averaged daily horizontal diffuse irradiation	32
2.3	Annual-averaged diffuse irradiation	34
2.4	Daily horizontal global irradiation	37
2.5	Daily horizontal diffuse irradiation	38
2.6	The inequality of the daily and monthly-averaged regressions	42
2.7	Daily slope irradiation	45
2.8	Exercises	49
	References	50

3 HOURLY HORIZONTAL IRRADIATION AND ILLUMINANCE — 53
3.1 Monthly-averaged hourly horizontal global irradiation — 53
3.2 Monthly-averaged hourly horizontal diffuse irradiation — 55
3.3 Hourly horizontal global irradiation — 59
3.4 Hourly horizontal diffuse irradiation — 71
3.5 Hourly horizontal illuminance — 76
3.6 Daylight factor — 89
3.7 Frequency distribution of irradiation — 92
3.8 Frequency distribution of illuminance — 96
3.9 Exercises — 100
References — 102

4 HOURLY SLOPE IRRADIATION AND ILLUMINANCE — 109
4.1 Slope beam irradiance and illuminance — 110
4.2 Sky clarity indices — 110
4.3 Sky-diffuse irradiance models — 111
4.4 Slope illuminance models — 133
4.5 Radiance and luminance distributions — 136
4.6 Luminance transmission through glazing — 147
4.7 Exercises — 150
References — 151

5 SOLAR SPECTRAL RADIATION *H. Kambezidis* — 155
5.1 Monochromatic solar spectral radiation — 155
5.2 Absorption and scattering of solar irradiance — 156
5.3 Spectral radiation model — 161
References — 163

6 GROUND ALBEDO — 165
6.1 Estimation of ground-reflected radiation — 166
6.2 Models for ground-reflected radiation — 169
6.3 Albedo atlas for the United Kingdom — 170
6.4 Estimation of monthly-averaged albedo — 171
References — 178

7 PSYCHROMETRICS — 181
7.1 Psychrometric properties — 181
7.2 Hourly temperature model — 184
References — 186

PROJECTS — 187

APPENDIX A: International daylight measurement programme — 189
APPENDIX B: Mean-monthly weather data for selected UK sites — 190
APPENDIX C: Instructions for use of FORTRAN programs — 192

INDEX — 193

FOREWORD

During the last decade there has been much research into solar radiation and daylighting in relation to environmental design. New data have been collected – particularly through the CIE/WMO Daylight Measurement Programme and its related projects – and new empirical models have been developed. Dr Muneer has been active in both aspects of the work.

Many numerical techniques now exist for calculating the distribution of radiation on and within buildings. This gives the designer considerable predictive power, but at the cost of maintaining knowledge of a large and changing literature. Published algorithms vary in scope, accuracy and length; in some cases several alternative procedures are available for estimating the same physical quantity.

The value of this book is that an expert in the subject has made a personal selection of applicable formulae, and presented them in a comprehensive and consistent format, both on paper and in the form of computer programs. Books such as this are indispensable references for the research worker and for the practising engineer.

Peter Tregenza
University of Sheffield

PREFACE

The aim of this work is to provide both a reference book and a text on solar radiation and daylight models. The book has grown out of the author's past 20 years of first hand experience of dealing with the relevant data from four continents: India, where the author grew up; the USA, where he got his advanced schooling; and Africa and the UK, where he taught and researched. Some of that work has been published in a series of technical articles. A concurrent and interesting activity in which the author is involved is the production of the new Chartered Institution of Building Services Engineers' *Guide for Weather and Solar Data*. This work provided an opportunity to liaise with colleagues from both sides of the Atlantic. The author was also fortunate to be awarded the Royal Academy of Engineering's fellowship to visit Japan on an extended study leave. Through this opportunity the author was able to examine the abundance of solar irradiance and illuminance data now being collected in the Far East. The models presented herein are applicable for a very wide range of locations world-wide, in particular in Europe, America, India and the Pacific Rim.

The text also emphasises the importance of good structure in the presentation of the computational algorithms. The chapters and sections have been divided in a manner which represents not only the chronological development of the knowledge base but also the algorithmic flow from a coarse to a more refined basis of calculation.

FORTRAN is one of the most widely used programming languages in engineering applications. A special feature of this text is that it includes 31 programs, provided in the *.For and *.Exe formats. The former format enables the user to make any changes such as providing data via prepared electronic files or to embed these routines in their own simulation or other programs. For example, the earlier work performed by the author involved liaison with the developers of ESP and SERI-RES building energy simulation packages to incorporate some of the enclosed solar radiation routines. The *.Exe files are for users who may not have access to FORTRAN compilers. These files may be run directly from the DOS level.

The enclosed suite of FORTRAN programs, contained in the compact disk, was designed and written by the author, based on several years of research and consultancy experience. The programs cover almost all aspects of solar radiation and daylight related computations. All programs included herein are introduced via examples and readers are encouraged to try them out as they progress through the book. Exercises as well as project work are additionally provided to enable further practice on the routines. To this end electronic files (data bases) with solar and other data are also included in the compact disk.

The following program copying policy is to be maintained. Copies are limited on a one-person/one-machine basis. Backup copies may be kept by each individual reader as required. However, further distribution of the programs constitutes a violation of the copy agreement and hence is illegal. Although the enclosed suite of computer programs has been prepared with great care and subjected to rigorous quality control checks, neither the author nor the publisher warrants the performance or the results that may be obtained by using the programs. Accordingly, the accompanying programs are licensed on an 'as-is' basis. Their performance or fitness for any particular purpose is assumed by the purchaser or user of this book.

The author welcomes suggestions for additions or improvements.

ACKNOWLEDGEMENTS

The author is indebted to the following individuals for their support: Alex Young, Peter Tregenza, Gurudeo Saluja, Hiroshi Nakamura, John Darby, Seaton Baxter, Bernard Yallop, Mike Holmes, Geoff Levermore, Ken Butcher, Eric Keeble, John Fulwood, Malcolm Lee, Phillip Haves, Paul Littlefair, Ian McCubbin, Phil Dolley, Kerr MacGregor, Maureen Kvebekk, Jean Lebrun, Costas Kaldis, Georges Liebecq, Werner Platzer, Fred Sick, Andreas Wagner, John Mardaljevic, Ghulam Dastgir, Roddy Angus, Naser Abodahab, Baolei Han, Gillian Weir, Mehreen Gul and Yasuko Koga.

The present text is the culmination of research undertaken by the author over the past two decades. Many organisations have either sponsored or actively supported the author's scholarly programme of work, and noteworthy among them are: the Scottish Education Department; Robert Gordon University; General Electric PLC; University College, Oxford; the Leverhulme Trust; Université de Liège, Belgium; the Royal Academy of Engineering, London; the Chartered Institution of Building Services Engineers, London; and the British Council through its offices in Germany and Greece. Their contributions are gratefully acknowledged.

The help extended by the publisher Neil Warnock-Smith is particularly appreciated. The author is grateful to George Pringle and Mandy Weir for their assistance. Above all the author would like to extend special thanks to his parents, wife and children for being extremely supportive throughout.

ELECTRONIC FILES ON COMPACT DISK

Computer programs

Prog1-1.For	Day number of the year for a given date
Prog1-2.For	Julian day number and day of the week for a given date
Prog1-3.For	Low precision algorithm for equation of time (EOT) and solar declination (DEC)
Prog1-4.For	Medium precision algorithm for EOT
Prog1-5.For	Medium precision algorithm for DEC
Prog1-6.For	High precision algorithm for EOT, DEC and solar geometry
Prog1-7.For	Sunrise, sunset and twilight times
Prog1-8.For	Distance between two locations
Prog2-1.For	Monthly-averaged horizontal global, diffuse and beam irradiation
Prog2-2.For	Daily horizontal global, diffuse and beam irradiation
Prog2-3.For	Monthly-averaged slope irradiation
Prog2-4.For	Daily slope irradiation
Prog3-1.For	Monthly-averaged hourly horizontal global, diffuse and beam irradiation
Prog3-2.For	Hourly horizontal global and diffuse irradiation using meteorological data
Prog3-3.For	Hourly diffuse fraction of horizontal global irradiation
Prog3-4.For	Horizontal global, diffuse and beam daylight illuminance, Perez et al. (1990) model
Prog3-5.For	Daylight factors for CIE overcast sky
Prog3-6.For	Frequency distribution of clearness index, US locations
Prog3-7.For	Frequency distribution of clearness index, tropical locations
Prog3-8.For	Frequency distribution of clearness index, UK and northern European locations
Prog3-9.For	Frequency distribution of daylight illuminance, Tregenza (1986) model
Prog4-1.For	Slope global, diffuse and beam irradiance
Prog4-2.For	Slope global, diffuse and beam illuminance
Prog4-3.For	Sky luminance distributions, relative co-ordinates

xiv   ELECTRONIC FILES ON COMPACT DISK

Prog4-4.For   Sky luminance distributions, absolute co-ordinates
Prog4-5.For   Incidence angle of luminance from a given sky patch
Prog4-6.For   Illuminance transmission functions for multiple glazed windows

Prog5-1.For   Spectral radiation model (SRM)

Prog7-1.For   Psychrometric properties, given dry-bulb and wet-bulb temperatures
Prog7-2.For   Psychrometric properties, given dry-bulb temperature and relative humidity
Prog7-3.For   Conversion from daily to hourly temperatures

Input files

In3-2.Csv   Sample input file for Prog3-2.For (must reside with Prog3-2.For)
In3-5.Csv   Data file required for executing Prog3-5.For (must reside with Prog3-5.For)
In4-3.Csv   Data file required for executing Prog4-3.For and Prog4-4.For (must reside with Prog4-3.For and Prog4-4.For)

Data files

File1-1.Csv   Solar declination angle and equation of time (must reside with Prog3-9.For)
File2-1a.Csv Mean-monthly meteorological data for the United Kingdom, January–June
File2-1b.Csv Mean-monthly meteorological data for the United Kingdom, July–December
File3-1.Csv   Five minute averaged measured solar data for Edinburgh, April 1993
File5-1.Csv   Measured spectral radiation data for Athens
File7-1.Csv   Dry-bulb temperature and humidity ratio data for psychrometric chart

FIGURES

1.5.1	Solar geometry of a sloped surface
1.6.1	Sun-path geometry for an approximate latitude of 50°N
1.8.1	Variation of daylight and twilight
1.11.1	Plot of residuals for evaluating the adequacy of the model
2.1.1	Calculation scheme for monthly-averaged daily sloped irradiation
2.1.2	Relationship between average clearness index and sunshine fraction
2.2.1	Variation of monthly-averaged diffuse ratio against clearness index
2.3.1	Variation of annual-averaged diffuse ratio against clearness index
2.4.1	Calculation scheme for daily sloped irradiation
2.5.1	Regression curves for daily diffuse ratio, Indian locations
2.5.2	Regression curves for daily diffuse ratio, UK locations
2.5.3	Regression curves for daily diffuse ratio, World-wide locations
2.7.1	Measured versus computed daily sloped irradiation for Easthampstead, UK (isotropic model)
2.7.2	Measured versus computed daily sloped irradiation for Easthampstead, UK (anisotropic model)
2.7.3	Measured versus computed daily sloped irradiation for Lerwick, UK (isotropic model)
2.7.4	Measured versus computed daily sloped irradiation for Lerwick, UK (anisotropic model)
3.1.1	Calculation scheme for monthly-averaged hourly sloped irradiation
3.1.2	Ratio of hourly to daily global irradiation
3.2.1	Ratio of hourly to daily diffuse irradiation
3.2.2	Individual (not averaged) values of r_D at 0.5 h from solar noon
3.2.3	Values of r_D at 0.5 h from solar noon for two fixed values of ω_s
3.3.1	Calculation scheme for hourly sloped irradiation
3.3.2	Evaluation of the MRM for clear skies
3.3.3	Evaluation of the MRM for overcast skies
3.3.4	Correlation between hourly diffuse and beam irradiation
3.3.5	Evaluation of the MRM for non-overcast skies
3.3.6	Performance of the MRM for daily irradiation
3.4.1	Hourly diffuse ratio versus clearness index for Camborne, UK
3.4.2	Hourly diffuse ratio versus clearness index for world-wide locations
3.5.1	Lighting control schematic

3.5.2	Performance of luminous efficacy models
3.5.3	Performance of average global luminous efficacy model
3.5.4	Performance of average diffuse luminous efficacy model
3.5.5	Relationship between global and diffuse luminous efficacy and clearness index at Fukuoka
3.5.6	Evaluation of Perez et al. model against Fukuoka data
3.6.1	Window schematic for Example 3.6.1
3.6.2	Availability of diffuse illuminance for world-wide locations
3.7.1	Frequency of occurrence of K_T for an Indian location
3.7.2	Individual K_T curves for Indian locations
3.7.3	Comparison of K_T curves for average clearness index 0.3
3.7.4	Comparison of K_T curves for average clearness index 0.5
3.7.5	Generalised K_T curves for USA
3.7.6	Generalised K_T curves for India
3.7.7	Generalised K_T curves for the UK
3.8.1	Frequency of occurrence of a given horizontal illuminance as a fraction of the mean illuminance
3.8.2	Derived cumulative distributions of global illuminance at Uccle for June and December
3.8.3	Standard working year daylight availability: cumulative global illuminance frequency
3.8.4	Standard working year daylight availability: cumulative diffuse illuminance frequency
3.8.5	Standard working year daylight availability: cumulative global illuminance frequency, London and Edinburgh
4.0.1	Luminance distribution for overcast skies
4.3.1	Solar geometry for an inclined surface
4.3.2	Relationship between shaded vertical and horizontal diffuse illuminance and irradiance at Chofu
4.3.3	Ratio of shaded vertical surface to horizontal diffuse incident energy
4.3.4	Relationship between vertical sun-facing background-sky diffuse illuminance and irradiance fraction and sky clarity at Fukuoka
4.3.5	Averaged background-sky diffuse illuminance and irradiance fraction versus sky clarity
4.3.6	Evaluation of slope irradiance models for a north-facing surface
4.3.7	Evaluation of slope irradiance models for an east-facing surface
4.3.8	Measured and estimated irradiance, north-facing surface
4.3.9	Measured and estimated irradiance, east-facing surface
4.3.10	Measured and estimated irradiance, south-facing surface
4.3.11	Measured and estimated irradiance, west-facing surface
4.5.1	Geometry of the sky elements for computation of luminance distribution
4.5.2	Detail of the sky patch shown in Figure 4.5.1
4.5.3	Microsoft Excel doughnut diagram for luminance distribution
4.6.1	Geometry of a given sky patch

FIGURES   xvii

5.3.1	Comparison of computed and measured spectral irradiance at Athens, Greece
6.1.1	Variation of albedo of bare soil and short grass with solar altitude
6.1.2	Variation of albedo of water surface and snow-covered ground with solar altitude and cloud cover
6.1.3	Effect of ageing of snow on albedo
6.1.4	Variation of albedo of a snow surface with accumulated temperature index since the last snowfall
6.3.1	Mean number of days with snow lying at 0900 GMT for November
6.3.2	Mean number of days with snow lying at 0900 GMT for December
6.3.3	Mean number of days with snow lying at 0900 GMT for January
6.3.4	Mean number of days with snow lying at 0900 GMT for February
6.3.5	Mean number of days with snow lying at 0900 GMT for March
6.3.6	Mean number of days with snow lying at 0900 GMT for April
7.1.1	Psychrometric chart based on Prog7-1.For
7.2.1	Evaluation of ASHRAE hourly temperature model for dry-bulb temperature
7.2.2	Evaluation of ASHRAE hourly temperature model for wet-bulb temperature

TABLES

1.2.1	Coefficients for Eq. (1.2.2)
1.2.2	Equation of time: accuracy evaluation for the 21st day of each month
1.2.3	Equation of time and solar declination angle for the year 2002
1.2.4	Accuracy evaluation of EOT models
1.4.1	Accuracy evaluation of DEC models
1.4.2	Solar declination: evaluation for the 21st day of each month
1.5.1	Solar geometry: evaluation for the 21st day of each month
1.10.1	CIE standard spectral relative sensitivity of the daylight adapted human eye
1.10.2	World Meteorological Organisation classification of pyranometers
1.11.1	Percentile values for Student's t-distribution
2.1.1	Monthly-averaged horizontal daily extraterrestrial irradiation
2.1.2	Coefficients for use in Eq. (2.1.1)
2.2.1	Monthly-averaged horizontal daily global and diffuse irradiation
2.3.1	Annual irradiation data for world-wide locations
2.7.1	Monthly-averaged daily irradiation for Easthampstead (51.383°N)
3.1.1	Monthly-averaged hourly horizontal irradiation for Eskdalemuir (55.3°N, 3.2°W)
3.3.1	Normal composition of clean atmosphere
3.3.2	Coefficients for use in Eq. (3.3.1)
3.3.3	Coefficients for Eqs (3.3.13)–(3.3.17)
3.3.4	Accuracy evaluation of hourly MRM, 1985–94 data
3.3.5	Data for Examples 3.3.1 and 3.4.1: London (51.5°N, 0.2°W), 14 April 1995
3.3.6	Evaluation of the MRM for monthly-averaged hourly irradiation
3.3.7	Performance of the MRM for monthly-averaged daily irradiation
3.5.1	Comparison of luminous efficacy models against Edinburgh data
3.5.2	Coefficients for Perez et al. luminous efficacy and zenith luminance model
3.5.3	Performance of Perez et al. and Littlefair luminous efficacy models, North London data
3.5.4	Input/output data for Example 3.5.1: North London data, 1 April 1992

TABLES

3.8.1	Frequency of occurrence of a given horizontal illuminance
3.9.1	Synoptic and radiation data for Easthampstead, UK (51.383°N), June 1991
4.3.1	Coefficients for Perez et al. slope irradiance and illuminance model
4.3.2	Evaluation of slope irradiance models at an hourly level, Edinburgh (55.95°N), August 1993
4.3.3	Measured and computed slope irradiation for Edinburgh (55.95°N), 12 August 1993
4.3.4	Evaluation of r_G, r_D, isotropic, Reindl et al. and Muneer models
4.4.1	Comparison of measured and computed illuminance for Edinburgh, 12 August 1993
4.5.1	Values of coefficient b in Eq. (4.5.1), overcast sky
4.5.2	Coefficients to be used in Eq. (4.5.5)
4.5.3	Measured luminance distribution data for intermediate sky, SOLALT=10 degrees
4.5.4	Measured luminance distribution data for intermediate sky, SOLALT=20 degrees
4.5.5	Measured luminance distribution data for intermediate sky, SOLALT=30 degrees
4.5.6	Measured luminance distribution data for intermediate sky, SOLALT=40 degrees
4.5.7	Measured luminance distribution data for intermediate sky, SOLALT=50 degrees
4.5.8	Measured luminance distribution data for intermediate sky, SOLALT=60 degrees
4.5.9	Measured luminance distribution data for intermediate sky, SOLALT=70 degrees
4.5.10	Coefficients for Perez et al. all-sky luminance distribution model
4.5.11	Comparison of Perez et al. luminance distribution model against measured data from Japan
4.5.12	Output for Example 4.6.3
5.2.1	Turbidity values for various visibilities
6.1.1	Albedo of soil covers
6.1.2	Albedo of vegetative covers
6.1.3	Albedo of natural surfaces
6.1.4	Albedo of building materials
7.2.1	Diurnal temperature swing

INTRODUCTION

Solar radiation and daylight are essential to life on earth. Solar radiation affects the earth's weather processes which determine the natural environment. Its presence at the earth's surface is necessary for the provision of food for mankind. Thus it is important to be able to understand the physics of solar radiation and daylight, and in particular to determine the amount of energy intercepted by the earth's surface. The understanding of the climatological study of radiation is comparatively new. Until 1960 there were only three stations in north-west Europe with irradiation records exceeding a 25 year period. In the United Kingdom it was only in the 1950s that Kipp solarimeters were installed by the Meteorological Office. Similarly, daylight was not recorded on a continuous basis and up to 1970 only seven sites across the UK measured horizontal illuminance. Furthermore, until 1992 there were no records of vertical illuminance for any location in the UK north of Watford (51.7°N) leaving the majority of the country without these measurements. At present vertical illuminance measurements exist for a short period and for only four sites across the country – Watford, Manchester, Sheffield and Edinburgh. These stations were operated in response to the call made by the Commission Internationale de l'Éclairage (CIE) under which 1991 was declared the International Daylight Measurement Year. Appendix A contains a list of daylight measurement stations world-wide.

Daylight is necessary for the production of all our agricultural produce and sustains the food chain through the process of photosynthesis. Photosynthesis is a biological phenomenon which describes the ability of plant life to convert light into chemical energy for growth. Daylight is one of the most important parts of the solar spectrum; it is the band of the sun's energy that we associate with day and night and has been the centre of much attention in recent years for a variety of reasons.

The initial research carried out by Ångström and others was concerned with the relationship between irradiation and sunshine duration. Since then research in this field has come a long way. The aim of this book is to further the understanding of the physics of short-wave irradiation, with particular emphasis on the development of mathematical models for computational purposes. The terms solar radiation, irradiation, radiance, irradiance, luminance and illuminance are frequently encountered in the literature and a note on their use is perhaps appropriate at this stage. Solar radiation (W/m^2) or luminance (lm/m^2 or lx) refers to the energy emanating from the sun. Luminance is the energy contained within the visible part of the solar radiation spectrum (0.39 to 0.78 μm). The terms irradiation ($W\,h/m^2$ or J/m^2) and illumination ($lm\,h/m^2$) refer to the cumulative energy incident on a surface in a given period of time. Irradiance and illuminance refer to the instantaneous incident energy. As would be expected, daylight and solar radiation possess similar physical characteristics and the modelling experience of one quantity helps the understanding of the other.

The interception of solar radiation by arbitrary surfaces is a function of their geometry and a determinant of their microclimatic interaction, i.e. the energy exchange between the surface and the surroundings. Estimation of horizontal irradiation is one task, assessment of insolation (solar irradiation) on slopes is another. Insolation availability of arbitrary sloped surfaces is a prerequisite in many sciences. For example, agricultural meteorology, photobiology, animal husbandry, daylighting, comfort air-conditioning, building sciences and solar energy utilisation all require insolation availability on slopes. In agricultural meteorology, the importance of net radiation in determining crop evaporation is well documented. It has been suggested that in the climate of the British Isles the annual enthalpy of evaporation from short grass is equal to the annual net radiation. A similar case occurs on a daily basis. Net radiation is also required in estimating the heating coefficient of a field, which is a key index for the soil germination temperature. The first step in determining net radiation is the incident short-wave radiation on the surface of the field, which may be situated on a slope.

Photosynthesis is an important phenomenon in photobiology. This term is commonly reserved for the process by which green plants are able to convert light into chemical energy. However, the absorption of energy is selective as far as the wavelength of the incident radiation is concerned. Therefore a spectral irradiation model is required in the applications of photobiology.

The effects of solar radiation are also of interest in the breeding of cattle, sheep and other livestock. It is usually the major factor limiting the distribution of stock in the tropics. The heat load on an animal is the result of solar irradiation and in some cases its magnitude could be several times the animal's normal heat production. Nature helps, however, in keeping down the heat load by having a low absorptivity of the animal's coat and by providing an insulating barrier in the form of thick fleece. Nevertheless, the limited ability of the animal to vaporise moisture, and thus regulate the dissipation of solar heat load, makes the effect of solar radiation on its surface an important factor. The problems of solar heat loads are not limited to the tropics. Even in Scotland the insolation levels may induce a considerable load in the summer season.

The rising cost of electricity has provided the motive for making best use of daylight. Utilisation of daylight and solar radiation has led to new architectural developments. Typical design elements include atria, sloping façades and large windows. But although there are new opportunities for making use of daylight, there is a need for comprehensive information on appropriate calculations. By incorporation of realistic prediction methods, daylight and passive solar design can provide a reduction in energy costs. The need for prediction methods for daylight is even more genuine owing to the fact that world-wide there is very limited measured illuminance data.

There is an increasing concern over our planet's 'global warming'. Since 1765 levels of greenhouse gases have increased substantially: CO_2 by 25%, CH_4 by 112% and nitrous oxides by 9%. The world-wide use of energy is rising by 2.5% a year, most of which is attributable to accelerated consumption in the developed countries. It has been estimated that, from a sustainability viewpoint, the developed countries will have to cut their use of energy by a factor of ten within a generation. Some proponents of solar energy are calling for a complete substitution of conventional sources of energy with renewables. Their thesis is that the use of fossil fuels for energy production, even in minor quantities, will merely postpone the collapse of the global environment.

Electricity production accounts for 39% of the UK's total CO_2 production. Electrical lighting in the UK accounts for an estimated 5% of the total primary energy consumed per annum. Exploitation of daylight can thus produce significant savings. Research has shown that savings of 20% to 40% are attainable for office buildings which utilise daylight effectively. The energy potential of daylight in the UK alone has been estimated at around one million tonnes of coal equivalent per annum by the year 2020.

The benefits and savings associated with daylight design are severalfold. Reduction of electrical lighting load due to the increased contribution of daylight results in lower sensible heat gains. This has the knock-on effect of lowering the cooling requirements of a building's air-conditioning. As cooling plants are high consumers of electricity, the costs associated with their operation can be as much as four times greater than those of heating. Furthermore the overall efficiency of a cooling plant is only 5% owing to the energy conversions associated with refrigeration and losses accumulating from electricity generation, transmission and final consumption. Thus, any reduction in electrical lighting load produces a much larger saving in primary energy consumption.

Buildings in the UK have traditionally been designed using daylight data recorded from the National Physical Laboratory in Teddington between 1933 and 1939. More recently new building constructions have employed illuminance data from Kew. The age of the data may not create any serious concern, although the Clean Air Acts have particularly affected major towns and cities across the country and could possibly influence present daylight levels. There is, however, concern centring around the lack of illuminance data from the major part of the country. In a relevant study it was shown that values of average horizontal illuminance in the northern part of the UK varied significantly from those reported for Kew; the differences were found to be as much as 32%. These differences have far reaching consequences for a building's performance.

As a consequence of the absence of adequate measured illuminance data, building designers have to rely on predictive tools and models. These models should be capable of accurately predicting illuminance values from meteorological parameters such as solar radiation. There are algorithms which allow the prediction of illuminance when solar irradiation is provided as an input parameter. Thus validated insolation models will provide information not only on the interception of irradiation, but also on daylight.

It has also been reported in the literature that during the past quarter-century many building air-conditioning systems were overdesigned. The resulting plant capacity, at least in the UK building stock, exceeds the true requirements by as much as 30%. This was due to the procedures used for load estimations. The World Energy Council has estimated that the overall efficiency of a vapour compression air-conditioning system is a mere 5%. Here, the overall plant efficiency is defined as the ratio of the energy extracted from the conditioned space to the energy consumed at the electrical generation plant. Thus, an overdesigned air-conditioning system imposes a serious penalty on the environment.

Two factors which may have contributed to the above overestimations are the assumption of isotropic diffuse irradiance for computing vertical surface energy gains and the use of hypothetical clear sky irradiation data. During the past decade several new and better solar radiation algorithms have evolved. These models indicate that the isotropic assumption overestimates the energy transmission through fenestration by as much as 40% for vertical surfaces in shade and over 20% for sun-facing surfaces under overcast conditions.

Modelling the availability of energy for the above mentioned applications requires knowledge of slope irradiation and illuminance on a monthly-averaged, daily (only applicable for irradiation) or hourly basis depending on how refined the analysis has to be. While modelling and simulation of energy systems would require determination of hourly horizontal and slope quantities, daily and monthly-averaged irradiation values would suffice for an abbreviated analysis. In the United Kingdom, for example, hourly diffuse and global irradiation on a horizontal plane are recorded by the Meteorological Office for 19 locations. Records for hourly global irradiation alone are available for a further seven stations. Additionally, daily global irradiation is recorded at 30 stations. For some of these locations the records exceed a period of 50 years. Slope irradiation measurements are, however, available for only two sites, Easthampstead (51.4°N) and Lerwick (60.2°N), which lie at the southern and northern extremes of the UK. Across western Europe as well as in the USA long-term records of slope irradiation are available for no more than a dozen locations. A similar situation exists in other parts of the world, e.g. in India only one and in China three stations presently log slope irradiance and illuminance.

The aim of this work is to address the relevant topics mentioned in the preceding discussion, and to provide those mathematical models and computer programs which enable computation of global, diffuse and beam irradiance and illuminance on arbitrary surfaces.

It was shown above that estimation of solar radiation and daylight is required in a number of scientific and engineering applications. However, there are only a handful of specialist books which address this subject with the required rigour. The past decade has seen a burst of activity in the relevant measurement and modelling spheres. As a consequence, previously available texts which deal with the subject of solar radiation and daylight models now appear dated. This book attempts to fill this gap. The present book is also different in its character, i.e. it presents a comprehensive suite of electronic programs which cover most aspects of solar radiation and daylight calculations. These programs are available on the accompanying PC-compatible compact disk. The programs are licensed on an 'as-is' basis.

With the dawn of the information technology revolution a significant proportion of what used to be printed material is now available in electronic format. In such publications the accompanying software enables the reader to resequence, reorganize and indeed create new material. The distinction between writer and reader is fast becoming blurred. As a matter of fact the experienced reader is able to push the boundaries of the electronic publication beyond the author's sphere of presentation. The medium and format of the present publication may be described as 'part soft, part hard', i.e. the text is available as hard copy while the accompanying algorithms and basic weather data are presented in electronic format. Electronic spreadsheets have come to be known as productivity software. However, at present the more sophisticated building energy simulation programs continue to use FORTRAN as the developmental medium. Therefore all algorithms included in this book have been written in the current version of that language.

The advantage of having computer programs in electronic format is that they may be copied and pasted in other larger simulation programs. This may be done effectively by making the present FORTRAN codes into subroutines or embedding them as part of the main body of the user's own programs. Of course, the programs may also be used as independent modules to evaluate any desired solar property or quantity on a one-off or iterative basis. The programs are being made available in both the *.For and *.Exe formats. The *.Exe files will enable those users who do not have access to a FORTRAN compiler to execute any of the accompanying routines in the DOS environment.

A further discussion on the present choice of the FORTRAN medium is provided at the beginning of the first chapter. It is possible, subject to adequate consumer demand, to provide all of the present algorithms in a spreadsheet medium (such as Excel or Lotus 1-2-3). Any comments from readers about improving the present format of the book or the accompanying FORTRAN programs would be appreciated.

1 FUNDAMENTALS

The past decade has seen a boom in the construction of energy efficient buildings which use solar architectural features to maximise the exploitation of daylight, solar heat and solar-driven ventilation. Daylight and building services engineers have therefore become accustomed to the use of specialist software for the physical simulation of buildings. Many of these software programs operate in the FORTRAN environment and, with new releases of FORTRAN available for the development work, this area of activity will see even more growth. Despite the fact that languages which are more structured than FORTRAN, e.g. C and Modula-2, are available, the former continues to be used widely in the engineering sector. Popular building physics simulation programs such as ESP and SERI-RES use the FORTRAN environment. 'The "Grand Challenges of Science and Engineering" for the foreseeable future will involve the use of FORTRAN-like programs on supercomputers' was an observation made in a recent report prepared for the President of the United States, as quoted by Edgar (1992). After its creation in the 1950s, FORTRAN developed many different dialects. However, the current standard is the ANSI X3.9-1978, widely known as FORTRAN 77. Currently, developments are under way for updating FORTRAN 77 to FORTRAN 90 and a good discussion of the add-on features is provided by Nyhoff and Leestma (1995). This book presents all routines in the current standard, i.e. FORTRAN 77.

The Commission Internationale de l'Éclairage (CIE) declared the year 1991 as the International Daylight Measurement Year. Consequently, new activity was seen across the globe on this front. This activity is called the International Daylight Measurement Programme (IDMP). In the UK as a result of the CIE call, horizontal and vertical illuminance and irradiation measurements have been carried out for four sites, i.e. Watford, Manchester, Sheffield and Edinburgh. These records are available on a minute-by-minute basis (Tregenza, 1994). Any sun position software which utilises data at this frequency warrants high accuracy. Thus older algorithms based on calculating the solar declination angle and the equation of time on a once a day basis will gradually become obsolete. It is worth noting that astronomical almanacs report values of the above variables for 0 hour Universal Time (UT).

In this chapter algorithm and software routines are presented which enable calculation of the sun's position and the related geometry. The present set of algorithms includes low to high accuracy models. The high precision algorithm for solar position calculation was developed by Yallop (1992), a leading astronomer. These algorithms are dynamic and take account of the change of the above mentioned parameters, year upon year. The validity period for the present high accuracy algorithm (Prog1-6.For and Prog1-6.Exe in the disk which accompanies this book) is from 1980 to 2050.

Disparate practice has been adopted by meteorological stations across the globe in measuring hourly solar radiation. While in the United Kingdom the irradiation data are available against apparent solar time (AST), many other countries use the local civil time (LCT) as the reference for all records. Under the CIE International Daylight Measurement Programme the illuminance was recorded world-wide against the local civil time. It is therefore necessary that appropriate algorithms are available for the conversion from one system to another. Basic concepts and definitions are introduced herein which are a prerequisite for obtaining the sun's position.

1.1 Solar day

A solar day is defined to be the interval of time from the moment the sun crosses the local meridian to the next time it crosses the same meridian. Owing to the fact that the earth rotates in a diurnal cycle as well as moves forward in its orbit, the time required for one full rotation of the earth is less than a solar day by about 4 minutes.

1.1.1 Day number

In many solar energy applications one needs to calculate the day number (DN) corresponding to a given date. DN is defined as the number of days elapsed in a given year up to a particular date. Examples of this application are the estimation of the equation of time and solar declination angle using low precision algorithms, and the extraterrestrial irradiance and illuminance at any given time. In the present section DN is obtained via Prog1-1.Exe. The FORTRAN routine for DN and the aforementioned .Exe file are available in the accompanying disk. Users may wish to incorporate the calculation of DN in their own FORTRAN routines and this would be facilitated by electronically copying Prog1-1.For in their programs.

Example 1.1.1

Calculate the day numbers (DN) corresponding to 1 March 1996 and 1 March 1997.

The output from Prog1-1.For is as follows:

DN = 61 (for 1 March 1996)
DN = 60 (for 1 March 1997)

Notice that the program takes account of the leap years.

1.1.2 Julian day number and day of the week

In many astronomical calculations it is often necessary to count the number of days elapsed since a predetermined reference date (fundamental epoch). By convention this date has been fixed as the Greenwich Mean noon of 1 January 4713 BC. The number of days elapsed from this epoch to any given date is called the Julian day number (JDN). The calculation for JDN involves six steps and these are given in Duffett-Smith (1988).

The estimation of the day of the week is easy once JDN is known. This involves a three-step algorithm which is also available in Duffett-Smith (1988). Modern control algorithms for energy efficient buildings, such as optimum start algorithm, may find the use for the day of the week routine to differentiate between weekdays and weekends.

Prog1-2.For and Prog1-2.Exe enable estimation of JDN and day of the week..

Example 1.1.2

Find JDN and the day of the week corresponding to 25 December 1996.

Prog1-2.Exe provides the following output:

JDN = 2 450 442.5
day of the week = Wednesday

It is worth noting that JDN begins at 12 h Universal Time (UT) and is therefore 0.5 day ahead of the civil time.

1.2 Equation of time

The solar day defined above varies in length throughout the year owing to:

(a) the tilt of the earth's axis with respect to the plane of the ecliptic containing the respective centres of the sun and the earth; and
(b) the angle swept out by the earth–sun vector during any given period of time, which depends upon the earth's position in its orbit.

Thus, the standard time (as recorded by clocks running at a constant speed) differs from the solar time. The difference between the standard time and the solar time is defined as the equation of time (EOT). EOT may be obtained as expressed by Woolf (1968):

$$EOT = 0.1236 \sin x - 0.0043 \cos x + 0.1538 \sin 2x + 0.0608 \cos 2x \qquad (1.2.1)$$

where $x = 360 (DN - 1)/365.242$ and $DN = 1$ for 1 January in any given year. EOT may also be obtained more precisely as presented by Lamm (1981):

$$EOT = \sum_{k=0}^{5} A_k \cos(2\pi k N / 365.25) + B_k \sin(2\pi k N / 365.25) \qquad (1.2.2)$$

where N is the day in the four-year cycle starting after the leap year. Prog1-2.For discussed earlier is especially useful for such calculations and is incorporated in the present algorithm. Values of the A_k and B_k coefficients are given in Table 1.2.1. In any non-leap year, EOT assumes the value of near zero for 0 h UT on 15 April, 13 June, 1 September and 25 December. A high precision routine for EOT is implicit in Yallop's algorithm which is presented in Section 1.4.

Table 1.2.1 Coefficients for Eq. (1.2.2)

k	$A_k \times 10^3$ (h) (h)	$B_k \times 10^3$ (h) (h)
0	0.2087	0.000 00
1	9.2869	−122.290 00
2	−52.2580	−156.980 00
3	−1.3077	−5.160 20
4	−2.1867	−2.982 30
5	−1.5100	−0.234 63

FORTRAN 77 routines for EOT based on Eqs (1.2.1), (1.2.2) and Yallop's algorithm are respectively presented in Prog1-3.For, Prog1-4.For and Prog1-6.For. These FORTRAN files and the corresponding machine executable files are also included in the accompanying PC-compatible disk. Table 1.2.2 presents the accuracy evaluation of the above EOT models, compared against the reference *Astronomical Phenomena* (1993).

It may be of interest to note that up to the early seventeenth century the equation of time, then known as Equation Naturales, could be computed with an accuracy of half a minute. In 1675 Flamsteed Street, the first Astronomer Royal in England, produced a technique to reduce the error in obtaining EOT to within 6 seconds. Today there are models available which enable EOT to be computed to an accuracy of 3 seconds over a range of 60 centuries, e.g. Hughes et al. (1989)!

Table 1.2.2 Equation of time: accuracy evaluation for the 21st day of each month

	Low: Eq. (1.2.1)		Medium: Eq. (1.2.2)		High: Yallop (1992)		*Astronomical Phenomena* (1993)	
	(min)	(s)	(min)	(s)	(min)	(s)	(min)	(s)
January	−10	56	−11	8	−11	13	−11	14
February	−13	57	−13	43	−13	41	−13	41
March	−7	40	−7	30	−7	19	−7	19
April	2	40	1	14	1	12	1	13
May	4	21	3	28	3	28	3	29
June	−1	30	−1	28	−1	38	−1	38
July	−6	20	−6	18	−6	21	−6	21
August	−3	17	−3	19	−3	13	−3	14
September	8	46	6	43	6	46	6	46
October	16	23	15	8	15	17	15	16
November	14	24	14	23	14	13	14	12
December	2	13	2	13	2	06	2	05

Table 1.2.3 *(see pp. 6–7)* presents precise values of EOT and solar declination angle for the year 2002. This table has been especially prepared for this book by the Royal Greenwich Observatory, Cambridge. The choice of this particular epoch was made owing to it being the mid-year in the first leap year cycle of the next millennium. The accuracy of this table is better than 1 s for EOT and 1 minute of arc for the solar declination angle. This table may be used as a reference for the next two decades. Typically, the variation of EOT, year on year, is around a tenth of a minute.

Example 1.2.1

Calculate the equation of time (EOT) for 2 February 1993 at 0 h Greenwich Mean Time.

Table 1.2.4 Accuracy evaluation of EOT models (2 February 1993)

Source	Accuracy	Program name	Program output (h)	EOT (min)
Woolf (1968)	Low	Prog1-3.For	−0.225 7	−13.5
Lamm (1981)	Medium	Prog1-4.For	−0.226 5	−13.6
Yallop (1992)	High	Prog1-6.For	−0.228 294 9	−13.7
Astronomical Phenomena (1993)	Reference			−13.7*

* Corresponds to 0 h UT.

Using the above mentioned FORTRAN routines, Table 1.2.4 may be generated which enables the necessary accuracy evaluation.

1.3 Apparent solar time

Solar time is the time to be used in all solar geometry calculations. It is necessary to apply the corrections due to the difference between the longitude of the given locality (LONG) and the longitude of the standard time meridian (LSM). This correction is needed in addition to the above mentioned equation of time. Thus apparent solar time (AST) is given by

$$\text{AST} = \text{standard time (local civil time)} + \text{EOT} \pm [(\text{LSM} - \text{LONG})/15] \qquad (1.3.1)$$

All terms in Eq. (1.3.1) are to be expressed in hours. The algebraic sign preceding the longitudinal correction terms contained in the square brackets should be inserted as positive for longitudes which lie east of LSM and vice versa. The LSM and LONG themselves have no sign associated with them.

SOLAR RADIATION AND DAYLIGHT MODELS

Table 1.2.3
Equation of time and solar declination angle for the year 2002 (all values for 0 h Universal Time)

		Equation of time		Declination			Equation of time		Declination			Equation of time		Declination			Equation of time		Declination
		(min)	(s)	(deg)	(min)		(min)	(s)	(deg)	(min)		(min)	(s)	(deg)	(min)		(min)	(s)	(deg)
Jan.	0	-2	49	-23	6	Feb. 15	-14	9	-12	49	Apr. 2	3	45	4	46	May 18	3	38	19
	1	3	17	23	2	16	14	6	12	28	3	3	27	5	9	19	3	36	19
	2	3	46	22	57	17	14	3	12	7	4	3	9	5	32	20	3	33	19
	3	4	14	22	52	18	13	58	11	46	5	2	52	5	55	21	3	29	20
	4	4	41	22	46	19	13	53	11	25	6	-2	34	6	18	22	3	25	20
	5	-5	8	-22	39	20	-13	48	-11	4	7	2	17	6	41	23	3	21	20
	6	5	35	22	32	21	13	41	10	42	8	2	1	7	3	24	3	16	20
	7	6	1	22	25	22	13	34	10	20	9	1	44	7	26	25	3	11	20
	8	6	27	22	18	23	13	26	9	58	10	1	28	7	48	26	3	5	21
	9	6	52	22	9	24	13	18	9	36	11	-1	12	8	10	27	2	58	21
	10	-7	17	-22	1	25	-13	9	-9	14	12	0	56	8	32	28	2	51	21
	11	7	42	21	52	26	12	59	8	52	13	0	40	8	54	29	2	44	21
	12	8	5	21	42	27	12	49	8	29	14	0	25	9	16	30	2	36	21
	13	8	28	21	33	28	12	39	8	7	15	0	10	9	37	31	2	28	21
	14	8	51	21	22	Mar. 1	12	27	7	44	16	0	4	9	59	June 1	2	19	22
	15	-9	13	-21	12	2	-12	16	-7	21	17	0	18	10	20	2	2	10	22
	16	9	34	21	1	3	12	3	6	58	18	0	32	10	41	3	2	0	22
	17	9	55	20	49	4	11	51	6	35	19	0	45	11	2	4	1	51	22
	18	10	14	20	37	5	11	37	6	12	20	0	58	11	23	5	1	40	22
	19	10	33	20	25	6	11	24	5	49	21	1	11	11	43	6	1	30	22
	20	-10	52	-20	12	7	-11	10	-5	26	22	1	23	12	4	7	1	19	22
	21	11	9	19	59	8	10	56	5	3	23	1	35	12	24	8	1	7	22
	22	11	26	19	46	9	10	41	4	39	24	1	46	12	44	9	0	56	22
	23	11	42	19	32	10	10	26	4	16	25	1	57	13	4	10	0	44	22
	24	11	57	19	18	11	10	10	3	52	26	2	7	13	23	11	0	32	22
	25	-12	12	-19	3	12	-9	55	-3	29	27	2	17	13	42	12	0	19	23
	26	12	25	18	49	13	9	39	3	5	28	2	27	14	2	13	0	7	23
	27	12	38	18	33	14	9	22	2	41	29	2	36	14	20	14	0	6	23
	28	12	50	18	18	15	9	6	2	18	30	2	44	14	39	15	0	18	23
	29	13	1	18	2	16	8	49	1	54	May 1	2	52	14	57	16	0	31	23
	30	-13	12	-17	46	17	-8	32	-1	30	2	2	59	15	15	17	0	44	23
	31	13	21	17	29	18	8	15	1	6	3	3	6	15	33	18	0	57	23
Feb.	1	13	30	17	13	19	7	57	0	43	4	3	12	15	51	19	1	10	23
	2	13	38	16	56	20	7	40	0	19	5	3	18	16	8	20	1	23	23
	3	13	45	16	38	21	7	22	0	5	6	3	23	16	25	21	-1	36	23
	4	-13	51	-16	20	22	-7	4	0	28	7	3	27	16	42	22	1	49	23
	5	13	57	16	2	23	6	46	0	52	8	3	31	16	59	23	2	2	23
	6	14	1	15	44	24	6	28	1	16	9	3	34	17	15	24	2	15	23
	7	14	5	15	26	25	6	10	1	39	10	3	37	17	31	25	2	28	23
	8	14	9	15	7	26	5	52	2	3	11	3	39	17	47	26	-2	41	23
	9	-14	11	-14	48	27	-5	33	2	26	12	3	41	18	2	27	2	53	23
	10	14	13	14	29	28	5	15	2	50	13	3	42	18	17	28	3	6	23
	11	14	14	14	9	29	4	57	3	13	14	3	42	18	32	29	3	18	23
	12	14	14	13	49	30	4	39	3	37	15	3	42	18	46	30	3	30	23
	13	14	13	13	29	31	4	21	4	0	16	3	41	19	0	July 1	-3	42	23
	14	-14	12	-13	9	Apr. 1	-4	3	4	23	17	3	40	19	14	2	-3	53	23

Source: prepared by the Royal Greenwich Observatory, Cambridge, United Kingdom.

FUNDAMENTALS

Date	Equation of time (min)	(s)	Declination (deg)	(min)	Date	Equation of time (min)	(s)	Declination (deg)	(min)	Date	Equation of time (min)	(s)	Declination (deg)	(min)	Date	Equation of time (min)	(s)	Declination (deg)	(min)
3	4	5	22	59	Aug 18	3	59	13	13	Oct. 3	10	46	3	48	Nov. 18	14	57	19	8
4	4	16	22	55	19	3	46	12	54	4	11	5	4	12	19	14	45	19	22
5	4	26	22	49	20	3	32	12	34	5	11	23	4	35	20	14	31	19	36
6	−4	37	22	44	21	−3	17	12	15	6	11	41	−4	58	21	14	17	−19	49
7	4	47	22	38	22	3	2	11	55	7	11	59	5	21	22	14	2	20	3
8	4	57	22	31	23	2	47	11	34	8	12	16	5	44	23	13	46	20	15
9	5	6	22	24	24	2	31	11	14	9	12	33	6	7	24	13	30	20	28
10	5	15	22	17	25	2	15	10	54	10	12	49	6	30	25	13	12	20	40
11	−5	24	22	9	26	−1	59	10	33	11	13	5	−6	52	26	12	54	−20	52
12	5	32	22	1	27	1	42	10	12	12	13	20	7	15	27	12	35	21	3
13	5	39	21	53	28	1	24	9	51	13	13	35	7	37	28	12	16	21	14
14	5	47	21	44	29	1	7	9	30	14	13	50	8	0	29	11	55	21	24
15	5	53	21	35	30	0	48	9	8	15	14	4	8	22	30	11	34	21	35
16	−5	59	21	26	31	0	30	8	47	16	14	17	−8	44	Dec. 1	11	12	−21	44
17	6	5	21	16	Sep. 1	0	11	8	25	17	14	30	9	6	2	10	50	21	54
18	6	10	21	6	2	0	8	8	4	18	14	42	9	28	3	10	27	22	2
19	6	15	20	55	3	0	27	7	42	19	14	54	9	50	4	10	3	22	11
20	6	19	20	44	3	0	27	7	42	20	15	5	10	12	5	9	39	22	19
21	−6	22	20	33	5	1	6	6	57	21	15	15	−10	33	6	9	14	−22	27
22	6	25	20	21	6	1	26	6	35	22	15	25	10	55	7	8	48	22	34
23	6	27	20	9	7	1	46	6	13	23	15	34	11	16	8	8	23	22	40
24	6	29	19	57	8	2	7	5	50	24	15	43	11	37	9	7	56	22	47
25	6	30	19	44	9	2	27	5	28	25	15	50	11	58	10	7	29	22	52
26	−6	30	19	31	10	2	48	5	5	26	15	57	−12	18	11	7	2	−22	58
27	6	30	19	18	11	3	9	4	42	27	16	4	12	39	12	6	34	23	3
28	6	29	19	4	12	3	30	4	20	28	16	9	12	59	13	6	7	23	7
29	6	28	18	50	13	3	51	3	57	29	16	14	13	19	14	5	38	23	11
30	6	26	18	36	14	4	12	3	34	30	16	18	13	39	15	5	10	23	15
31	−6	24	18	22	15	4	33	3	11	31	16	21	−13	59	16	4	41	−23	18
1	6	21	18	7	16	4	55	2	48	Nov. 1	16	23	14	18	17	4	12	23	20
2	6	17	17	52	17	5	16	2	24	2	16	25	14	37	18	3	43	23	23
3	6	13	17	36	18	5	38	2	1	3	16	26	14	56	19	3	13	23	24
4	6	8	17	21	19	5	59	1	38	4	16	26	15	15	20	2	44	23	25
5	−6	2	17	5	20	6	21	1	15	5	16	25	−15	33	21	2	14	−23	26
6	5	56	16	48	21	6	42	0	51	6	16	23	15	52	22	1	44	23	26
7	5	50	16	32	22	7	3	0	28	7	16	20	16	10	23	1	14	23	26
8	5	43	16	15	23	7	25	0	5	8	16	17	16	27	24	0	45	23	25
9	5	35	15	58	24	7	46	0	19	9	16	13	16	45	25	0	15	23	24
10	−5	27	15	41	25	8	7	0	42	10	16	8	−17	2	26	0	15	−23	23
11	5	18	15	23	26	8	27	1	5	11	16	2	17	19	27	0	45	23	21
12	5	8	15	5	27	8	48	1	29	12	15	55	17	35	28	1	14	23	18
13	4	58	14	47	28	9	8	1	52	13	15	47	17	51	29	1	44	23	15
14	4	47	14	29	29	9	29	2	15	14	15	39	18	7	30	2	13	23	12
15	−4	36	14	10	30	9	48	−2	39	15	15	30	−18	23	31	−2	42	−23	8
16	−4	24	13	52	Oct. 1	10	8	−3	2	16	15	20	−18	38					
17	4	12	13	33	2	10	27	3	25	17	15	9	18	53					

8   SOLAR RADIATION AND DAYLIGHT MODELS

Example 1.3.1

What is the AST for Madison, Wisconsin corresponding to 10:30 a.m. local civil time on 2 February 1993? The following data are given: latitude = 43°N, longitude (LONG) = 89.4°W and longitude of standard time meridian (LSM) = 90°W.

Using Prog1-6.For, AST = 10.310 35 hours (10:19 a.m.). This value is in agreement with the solution given in Example 1.5.1 of Duffie and Beckman (1980).

1.4 Solar declination

The angle between the earth–sun vector and the equatorial plane is called the solar declination angle (DEC). As an adopted convention DEC is considered to be positive when the earth–sun vector lies northwards of the equatorial plane. Declination may also be defined as the angular position of the sun at noon (apparent solar time) with respect to the equatorial plane.

DEC may be obtained as expressed by Boes and reported in Kreider and Kreith (1981):

$$DEC = \sin^{-1} \{0.397\,95 \cos [0.985\,63 \,(DN - 1)]\} \tag{1.4.1}$$

Alternatively, it may be computed, with a medium to high accuracy, using the algorithm presented by Duffett-Smith (1988). This algorithm involves the following steps:

Step 1   The start of epoch for this algorithm is fixed as 0.0 January 1990. For any given date (either before or after the above epoch) calculate the number of days D passed since the start of the epoch.

Step 2   Calculate $N = 360D / 365.242\,191$: add or subtract multiples of 360 until N lies in the range 0–360 degrees.

Step 3   Find $M = N + \varepsilon_g - \omega_g$: If the result is negative, add 360 degrees. Here ε_g and ω_g are respectively the ecliptic longitude of the sun at epoch (= 279.403 303 degrees) and the ecliptic longitude of perigee (= 282.768 422 degrees).

Step 4   Find $E_c = (360/\pi)\, e \sin M$, where e is the eccentricity of orbit (= 0.016 713).

Step 5   Find $\lambda = N + E_c + \varepsilon_g$: if the result is more than 360 degrees, subtract 360.

Step 6   DEC = $\sin^{-1} (\sin \varepsilon \sin \lambda)$, where ε = 23.441 884 degrees is the obliquity of the ecliptic, the angle between the planes of the equator and the ecliptic.

Yallop's (1992) algorithm enables a high precision computation of EOT and DEC. The present routine is valid for the period 1980–2050 and has an accuracy of 3 s for EOT and 1 minute of arc for DEC. For a given year y, month m, day D, hour h, minute *min* and second s,

$$t = \{(UT/24) + D + [30.6m + 0.5] + [365.25(y - 1976)] - 8707.5\}/365\,25 \quad (1.4.2)$$

where Universal Time $UT = h + (min/60) + (s/3600)$. In Eq. (1.4.2), if $m > 2$ then $y = y$ and $m = m - 3$; otherwise $y = y - 1$ and $m = m + 9$. In the expression, $[x]$ denotes the integer part of x.

The following terms are then determined. If necessary add or subtract multiples of 360 degrees to G, L and GHA to set them in the range 0 to 360 degrees.

$G = 357.528 + 35\,999.05t$
$C = 1.915 \sin G + 0.020 \sin 2G$
$L = 280.460 + 36\,000.770t + C$
$\alpha = L - 2.466 \sin 2L + 0.053 \sin 4L$
Greenwich Hour Angle $GHA = 15UT - 180 - C + L - \alpha$
If necessary add or subtract multiples of 360 degrees to G, L and GHA to set them in the range 0 to 360 degrees.
obliquity of the ecliptic $\varepsilon = 23.4393 - 0.013t$

solar declination angle $DEC = \tan^{-1}(\tan \varepsilon \sin \alpha)$
equation of time $EOT = (L - C - \alpha) / 15$

FORTRAN 77 routines for DEC based on the Eq. (1.4.1), Duffett-Smith (1988) and Yallop (1992) algorithms are respectively presented in Prog1-3.For, Prog1-5.For and Prog1-6.For. These FORTRAN files and the corresponding machine executable files are also included in the accompanying disk. The electronic data file File1-1.Csv contains daily values of EOT and DEC computed for noon GMT. Where high computing speed is required this file may be used with the user's own routines.

Example 1.4.1

Calculate solar declination DEC for Madison, Wisconsin corresponding to 10:30 a.m. local civil time on 2 February 1993. Compare the results obtained using the algorithms due to Boes (Kreider and Kreith, 1981), Duffett-Smith (1988) and Yallop (1992).

Table 1.4.1 shows the results for accuracy evaluations. Table 1.4.2 compares the DEC

Table 1.4.1 Accuracy evaluation of DEC models (2 February 1993)

Source	Accuracy	Program name (h)	Program output (deg)
Kreider and Kreith (1981)	Low	Prog1-3.For	−17.1983
Duffett-Smith (1988)	Medium	Prog1-5.For	−16.8748
Yallop (1992)	High	Prog1-6.For	−16.8698
Astronomical Phenomena (1993)	Reference		−16.8667[*]

[*] Corresponds to 0 h UT.

Table 1.4.2 Solar declination: evaluation for the 21st day of each month (0 h GMT)

Accuracy	Low: Eq.(1.4.1)		Medium: Duffett-Smith (1988)		High: Yallop (1992)		Astronomical Phenomena (1993)	
	(deg)	(min)	(deg)	(min)	(deg)	(min)	(deg)	(min)
January	−21	53	−19	57	−19	57	−19	57
February	−12	47	−10	38	−10	38	−10	38
March	−1	20	0	9	0	9	0	9
April	11	5	11	48	11	47	11	47
May	19	50	20	9	20	8	20	8
June	23	27	23	26	23	26	23	26
July	20	27	20	31	20	31	20	31
August	11	47	12	11	12	11	12	11
September	0	7	0	47	0	47	0	47
October	−12	47	−10	38	−10	37	−10	37
November	−21	53	−19	52	−19	52	−19	52
December	−24	33	−23	26	−23	26	−23	26

estimates obtained via the three models, and compares them against the *Astronomical Phenomena* (1993).

As stated above, Table 1.2.3 provides high accuracy values of DEC. Year on year the variation of DEC is around a tenth of a degree of arc.

1.5 Solar altitude, azimuth and incidence angle

The sun's position in the sky can be described in terms of two angles: SOLALT, the elevation angle above the horizon; and SOLAZM, the azimuth from north of the sun's beam projection on the horizontal plane (clockwise = positive). These co-ordinates, which describe the sun's position, are dependent on AST, the latitude (LAT) and longitude (LONG) of the location, and DEC. The solar geometry may now be obtained from the following equations:

$$\sin \text{SOLALT} = \sin \text{LAT} \sin \text{DEC} - \cos \text{LAT} \cos \text{DEC} \cos \text{GHA} \quad (1.5.1)$$

$$\cos \text{SOLAZM} = \frac{\cos \text{DEC} (\cos \text{LAT} \tan \text{DEC} + \sin \text{LAT} \cos \text{GHA})}{\cos \text{SOLALT}} \quad (1.5.2)$$

The angle of incidence, INC at which the sun's beam strikes a sloped surface of any given tilt (TLT) can then be calculated from the solar altitude and azimuth and the orientation of the surface as expressed by its wall azimuth angle (WAZ). The sign convention adopted for WAZ is the same as that used for SOLAZM, i.e. clockwise from north is considered positive.

$$\text{INC} = \cos^{-1} [\cos \text{SOLALT} \cos(\text{SOLAZM} - \text{WAZ}) \sin \text{TLT} + \cos \text{TLT} \sin \text{SOLALT}] \quad (1.5.3)$$

FUNDAMENTALS 11

Table 1.5.1 Solar geometry: evaluation for the 21st day of each month (11 h apparent solar time for 40°N)

	Solar altitude		Solar azimuth*	
	Prog1-6 (deg)	Kreider and Kreith (1981) (deg)	Prog1-6 (deg)	Kreider and Kreith (1981) (deg)
January	28.5	28.0	163.8	164.0
February	37.6	37.0	161.1	161.0
March	48.0	48.0	157.0	157.0
April	59.1	59.0	151.2	151.0
May	66.5	66.0	142.8	143.0
June	68.9	69.0	137.2	138.0
July	66.6	66.0	142.2	143.0
August	59.0	59.0	150.5	151.0
September	48.3	48.0	157.0	157.0
October	37.3	37.0	161.2	161.0
November	28.4	28.0	164.0	164.0
December	25.1	25.0	165.3	165.0

* Clockwise from true north.

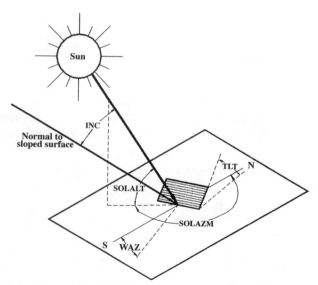

Figure 1.5.1 *Solar geometry of a sloped surface*

Figure 1.5.1 shows the angles relevant to the determination of the sun's position and the geometry for a tilted surface. Table 1.5.1 compares the computations of the present routine against Kreider and Kreith (1981). The maximum difference between presently computed values of solar altitude and azimuth and those given by Kreider and Kreith (1981) may be up to 0.5 degrees. The presently generated values of the solar co-ordinates are, of course, much more precise and the approximate nature of the older algorithms has thus been demonstrated. This is owing to the fact that the older algorithms use values of EOT and DEC for 0 hours UT, whereas the present routine evaluates these for the precise moment at which the solar geometry is required.

Example 1.5.1

Calculate the solar altitude and the sun's azimuth for Edinburgh, UK (lat. = 55.95°N, long. = 3.20°W) at 12 noon LCT on 21 March 1997. Also, find the angle of incidence of the sun's beam on a surface with a given tilt of 45 degrees and orientation 15 degrees west of south.

Note that the required input value of the wall azimuth angle WAZ is 195 degrees (= 180 + 15).

For the above data, Prog1-6.Exe produces the following output:

EOT = – 0.1196 h
AST = 11.667 h
solar hour angle = – 4.99 degrees
DEC = 0.3626 degrees
SOLALT = 34.27 degrees
SOLAZM = 173.95 degrees
INC = 19.35 degrees

The point of interest here is that SOLAZM is almost 6 degrees away from the usually assumed value of 180 degrees for solar noon LCT. For example, the solar geometry tables presented in *CIBSE Guide A2* (CIBSE, 1982) do not draw a line of demarcation

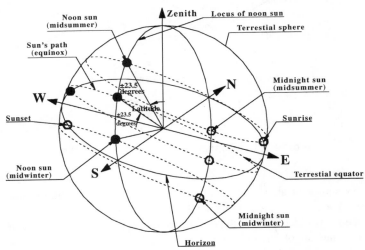

Figure 1.6.1 *Sun-path geometry for an approximate latitude of 50°N*

between the two time frames, LCT and AST. It has been demonstrated herein that this may lead to errors of the above mentioned magnitude.

1.6 Astronomical sunrise and sunset

Figure 1.6.1 shows the sun path geometry for a location with an approximate latitude of 50°N. Astronomers define sunrise and sunset as the moments at which the centre of the solar disk is along the horizon of the earth. It was shown above that the sun's position in the sky can be determined in terms of the elevation angle (SOLALT) and the azimuth of the sun's beam from north (SOLAZM). Using Eq. (1.5.1), the sunrise/sunset instance may be computed by setting SOLALT = 0. The sunrise/sunset instance may then be estimated by using Eq. (1.3.1) to obtain the standard time (local civil time).

1.7 Actual sunrise and sunset

The actual sunrise and sunset do not occur at the time when the sun's elevation is zero. This is due to the refraction of light by the terrestrial atmosphere. A ray of light travelling in vacuum from the sun where it is actually below the earth's horizon is bent towards the earth by the heavier medium, air (the average refractive index of atmospheric air is 1.0003). Hence actual sunrise appears slightly before astronomical sunrise and actual sunset occurs after astronomical sunset. Further, for locations which are higher than sea level, the sun will appear in the morning slightly earlier. Corrections have therefore to be made for the above refraction and altitude effects. These are expressed via the following equation for SOLALT which refers to the instance of actual sunrise or sunset:

$$SOLALT = -0.8333 - 0.0347\, H^{0.5} \tag{1.7.1}$$

In this equation H is to be given in metres above sea level (m ASL). Eq. (1.5.1) is then solved in conjunction with Eq. (1.3.1) to obtain the corresponding local civil time.

1.8 Twilight

Twilight is defined as the pre-sunrise or post-sunset period of partial daylight and is caused by the reflection and scattering of sunlight towards the horizon of any terrestrial observer. Twilight has an important biological and socio-religious significance. At any one instant, the twilight zone covers 20–25% of the surface area of our globe and humans, on average, live under the twilight band for a quarter of the time. In the tropics, owing to the sun's steep descent towards the horizon, twilight occupies only 10–15% of the diurnal cycle. However, in the higher latitudes such as those of northern Europe, twilight occupies up to two-fifths of the annual cycle.

Soon after sunset the illuminance progressively diminishes in an exponential manner until the sun sinks to an elevation angle of −18 degrees. This is the instance of the last stage of receipt of light emanating from the sun (astronomical twilight). The negative elevation angle, which corresponds to the period of twilight, is also expressed as the angle of depression. Thus, when SOLALT = -18 degrees, the depression angle = 18 degrees.

Various stages of twilight have been standardised, e.g. civil and nautical twilight, respectively, when the solar depression angles are 6 degrees and 12 degrees (see Figure 1.8.1). Civil twilight is the stage when enough illuminance exists to enable outdoor civil activity to continue unhindered without resorting to the use of electric street lighting. Nautical twilight is the stage which establishes the limit of the visibility of ships approaching a harbour.

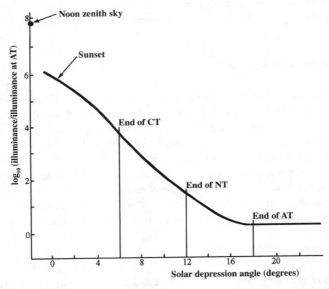

Figure 1.8.1 *Variation of daylight and twilight (AT, astronomical twilight; NT, nautical twilight; CT, Civil twilight)*

The following illuminance figures have been attributed to the famous physicist Kimball. For a horizontal surface under a cloudless sky:

sun at zenith:	103 000	lux
sun at horizon:	355	lux
end of civil twilight:	4.3	lux
end of astronomical twilight:	0.001	lux
full moon at zenith:	0.215	lux

It is with some fascination that we note that the change in illuminance from noon to astronomical twilight is around 100 million! Even between sunset and the end of twilight

FUNDAMENTALS   15

the change is about 400 000. Yet the human eye, being a logarithmic sensor, is able to cope with such wide ranging illuminance levels. Laboratory measurements of daylight and twilight are undertaken with sensors which operate in much narrower ranges.

An exhaustive review of the research undertaken on the twilight phenomenon and its measurement has been presented by Rozenberg (1966). Here, Prog1-7.Exe enables computation of the actual sunrise and sunset times and the three stages of twilight, described above. This is demonstrated via the following example.

Example 1.8.1

Calculate for the city of Edinburgh (lat. = 55.95°N, long = 3.20°W, altitude = 35 m ASL) the sunrise and sunset times and the times for the appearance of civil, nautical and astronomical twilight for 1 March 1996.

Prog1-7.Exe produces the following output:

sunrise at	0702 h
sunset at	1748 h
astronomical twilight appears at	0500 h
nautical twilight appears at	0543 h
civil twilight appears at	0626 h

It is worth noting that at midsummer, between the Arctic Circle and 48.5 degrees north, there is a belt with no true night, i.e. the astronomical twilight extends from sunset to sunrise. This is due to the fact that the solar depression angle for 21 June for 48.5°N (and even more so for the northerly latitudes) is less than 18 degrees, the last stage of astronomical twilight. Thus, the temporal period of no true night progressively widens as one crosses to more northerly latitudes. Prog1-7.Exe has the capability to handle such cases with robustness. The reader is invited to use this particular software for any of the northern locations for dates around midsummer, e.g. estimation of the data obtained in Example 1.8.1 for London for, say, 15 June.

1.9 Distance between two locations

In Chapter 2 of this book it will be shown that daily values of solar irradiation are more accurately estimated from local sunshine observations than by assignment from nearby pyranometric stations if the latter are more than 20 km away. For monthly-averaged daily totals this critical distance may be taken to be about 30 km. Thus, an algorithm is required which provides the distance between any two locations, pyranometric and candidate sites, to evaluate the applicability of the above constraints. The distance sought in this case is based on the principle of great circle navigation. The algorithm given below provides the distance and heading (for shortest flight path) from the source to the destination.

If we denote the geographical latitudes of the source and destination, respectively, as

LatS and LatD and the longitudes as LongS and LongD, then the distance between the locations is obtainable from the following suite of equations:

$$\text{Dterm} = [\sin(\text{LatS}) \sin(\text{LatD}) + \cos(\text{LongS}) \cos(\text{LongD}) \cos(\text{LongD} - \text{LongS})] \quad (1.9.1)$$
$$\text{distance (miles)} = 60 \, (180/\pi) \cos^{-1}(\text{Dterm}) \quad (1.9.2)$$
$$\text{distance (km)} = \text{distance (miles)} \times 1.609\,344 \quad (1.9.3)$$
$$\text{Hterm1} = (180/\pi) \cos^{-1}[\{\sin(\text{LatD}) - \sin(\text{LatS})\}/\{\cos(\text{LatS})(1 - \text{Dterm}^2)^{0.5}\}] \quad (1.9.4)$$
$$\text{Hterm2} = \sin(\text{LongS} - \text{LongD}) \quad (1.9.5)$$
$$\text{heading} = \text{Hterm1}, \qquad \text{Hterm2} > 0 \quad (1.9.6a)$$
$$\qquad\qquad = 360 - \text{Hterm1}, \quad \text{Hterm2} < 0 \quad (1.9.6b)$$

A program for the above computational scheme is given in Prog1-8.For. The following example demonstrates its use.

Example 1.9.1

An hour-by-hour energy simulation is to be carried out for a prospective office building, to be situated in the city of Edinburgh (lat. = 55.95°N, long. = 3.20°W). While other climatological data are available from Turnhouse (local airport), solar radiation records are available only for Mylnefield, Dundee (56.45°N, 3.067°W). Calculate the shortest distance between the two sites in order to investigate the feasibility of using Mylnefield solar data for Edinburgh.

The output from Prog1-8.Exe is as follows:

distance = 48.8 km
heading = 10.7 degrees (azimuth direction, clockwise from north)

Since the above distance is more than 30 km, neither the daily nor the monthly-averaged solar radiation data from Mylnefield may reliably be used for Edinburgh.

1.10 Solar radiation and daylight measurement

Routine measurement of diffuse solar energy from the sky and the global (total) radiation incident on a horizontal surface is usually undertaken by an agency such as the national meteorological office. For this purpose the measurement network uses pyranometers, solarimeters or actinographs. Direct or beam irradiation is measured by a pyrheliometer with a fast-response multijunction thermopile placed inside a narrow cavity tube. The aperture is designed such that it admits a cone of full angle around 6 degrees. Most of the irradiance sensors used across Europe are manufactured by Kipp and Zonen, while Eppley and Eko instruments are more widely used in the US and Japan respectively.

The CM 11 is regarded as the standard reference pyranometer owing to its accuracy, stability and quality of construction. The sensing element consists of a thermal detector which responds to the total power absorbed without being selective to the spectral distribution of radiation. The heat energy generated by the absorption of radiation on the black disk flows through a thermal resistance to the heat sink. The resultant temperature difference across the thermal resistance of the disk is converted into a voltage which can be read by computer. The double glass construction minimises temperature fluctuations from the natural elements and reduces thermal radiation losses to the atmosphere. The glass domes can collect debris over time and weekly cleaning is recommended. Moisture is prevented by the presence of silica gel crystals in the body of the CM 11. The pyranometers have a spectral response of between 335 and 2200 nm of the solar spectrum which includes the visible wavelength band.

Diffuse irradiance is measured by placing a shadow band over a pyranometer, adjustment of which is required periodically. Coulson (1975) provides an excellent account of these adjustments and the associated measurement errors for the above sensors, a brief summary of which is provided herein. The object of this section is to familiarise the reader with the order of errors encountered in the measurement of solar radiation and daylight. This is very much needed to establish the upper limit of refinement in the modelling work, i.e. no model can surpass the accuracy of the measured quantity. The present state of solar radiation and daylight models is such that they are approaching the accuracy limits set by the measuring equipment (Perez et al., 1990).

Radiation in the visible region of the spectrum is often evaluated with respect to its visual sensation effect on the human eye. The Commission Internationale de l'Éclairage (CIE) meeting in 1924 resulted in the adoption of a standard of the above wavelength dependent sensitivity. The CIE standardised sensitivity of the daylight adapted human eye is presented in Table 1.10.1.

Illuminance measuring sensors are known as either photometers or daylight sensors. These sensors are similar in construction to the above mentioned pyranometers. Daylight

Table 1.10.1 CIE standard spectral relative sensitivity of the daylight adapted human eye

Wavelength (μm)	Relative sensitivity	Wavelength (μm)	Relative sensitivity	Wavelength (μm)	Relative sensitivity
0.38	0.0000	0.51	0.5030	0.64	0.1750
0.39	0.0001	0.52	0.7100	0.65	0.1070
0.40	0.0004	0.53	0.8620	0.66	0.0610
0.41	0.0012	0.54	0.9540	0.67	0.0320
0.42	0.0040	0.55	0.9950	0.68	0.0170
0.43	0.0116	0.56	0.9950	0.69	0.0082
0.44	0.0230	0.57	0.9520	0.70	0.0041
0.45	0.0380	0.58	0.8700	0.71	0.0021
0.46	0.0600	0.59	0.7570	0.72	0.0010
0.47	0.0910	0.60	0.6310	0.73	0.0005
0.48	0.1390	0.61	0.5030	0.74	0.0003
0.49	0.2080	0.62	0.3810	0.75	0.0001
0.50	0.3230	0.63	0.2650	0.76	0.0001

illuminance is not one of the quantities routinely measured by the meteorological networks. There is therefore a dearth of measured illuminance data. However in 1991, in response to the call made by the CIE's International Daylight Measurement Programme, new activity was seen across the globe. For example during 1991–95 a total of 14 daylight and solar radiation measurement stations were operational in Japan. In France, the UK and the USA the respective number of stations were five, four and three. Most of the European sensors in current use were manufactured by the German vendors PRC Krochmann while the Japanese equipment has been built by the Tokyo-based Eko company.

Owing to the fact that there is a relative abundance of solar radiation data it is common practice to correlate the illuminance values against irradiation figures. Such models are known as luminous efficacy models and will be described in Chapter 3.

In many countries, the diurnal duration of bright sunshine is measured at a large number of places. The hours of bright sunshine are the time during which the sun's disk is visible. For over a century these data have been measured with the well known Campbell-Stokes sunshine recorder which uses a solid glass spherical lens to burn a trace of the sun on a treated paper, the trace being produced whenever the beam irradiation is above a critical level. Although the critical threshold varies loosely with the prevailing ambient conditions, the sunshine recorder is an economic and robust device and hence is used widely.

The limitations of the Campbell-Stokes sunshine recorder are well known and have been discussed in the *Observers' Handbook* (1969), Painter (1981) and Rawlins (1984). Some of the associated limitations with this device are that the recorder does not register a burn on the card below a certain level of incident radiation (about 150 to 300 W/m^2). On a clear day with a cloudless sky the burn does not start until 15–30 minutes after sunrise and usually ceases about the same period before sunset. This period varies with the season. On the other hand under periods of intermittent bright sunshine the burn spreads. The diameter of the sun's image formed by the spherical lens is only about 0.7 mm. However, a few seconds' exposure to bright sunshine may produce a burnt width of about 2 mm. Consequently, intermittent sunshine may be indistinguishable from a longer period of continuous sunshine.

A more sophisticated photoelectric sunshine recorder, called the Foster sunshine switch (Foster and Foskett, 1953), is in use by the US Weather Service. This device incorporates two photovoltaic cells, one shaded and the other exposed to the solar beam. Incident beam irradiation above a given threshold produces a differential output from the two cells, the diurnal duration of which determines the hours of bright sunshine.

1.10.1 Equipment error and uncertainty

With any measurement there exist errors, of which some are systematic and others are inherent in the equipment employed. Angus (1995) has provided an account of the measurement errors associated with solar irradiance and illuminance measurements. These are summarised herein. The most common sources of error arise from the sensors and their construction. These are broken down into the most general types of error as follows:

(a) cosine response
(b) azimuth response
(c) temperature response
(d) spectral selectivity
(e) stability
(f) non-linearity.

To be classed as a secondary standard instrument (such as the CM 11) pyranometers have to meet the specifications set out by the World Meteorological Organisation (WMO).

Of all the aforementioned errors, the cosine effect is the most apparent and widely recognised. This is the sensor's response to the angle at which radiation strikes the sensing area. The more acute the angle of the sun, i.e. at sunrise and sunset, the greater this error (at altitude angles of sun below 6 degrees). Cosine error is typically dealt with through the exclusion of the recorded data at sunrise and sunset times.

The azimuth error is a result of imperfections of the glass domes, and in the case of solarimeters the angular reflection properties of the black paint. This is an inherent manufacturing error which yields a similar percentage error as the cosine effect.

Like the azimuth error, the temperature response of the sensor is an individual fault for each cell. The photometers are thermostatically controlled, and hence the percentage error due to fluctuations in the sensor's temperature is reduced. However, the CM 11 pyranometers have a much less elaborate temperature control system. The pyranometers rely on the two glass domes to prevent large temperature swings.

Table 1.10.2 World Meteorological Organisation classification of pyranometers

Characteristic	Secondary standard	First class	Second class
Resolution (smallest detectable change in (W/m^2)	± 1	± 5	± 10
Stability (percentage of full scale, change/year)	± 1	± 2	± 5
Cosine response (percentage deviation from ideal at 10° solar elevation on a clear day)	< ± 3	< ± 7	< ± 15
Azimuth response (percentage deviation from ideal at 10° solar elevation on a clear day)	< ± 3	< ± 5	< ± 10
Temperature response (percentage maximum error due to change of ambient temperature within the operating range)	± 1	± 2	± 5
Non-linearity (percentage of full scale)	± 0.5	± 2	± 5
Spectral sensitivity (percentage deviation from mean absorptance 0.3 to 3 μm)	± 2	± 5	± 10
Response time (99% response)	< 25 s	< 1 min	< 4 min

The spectral selectivity of the CM 11 is dependent on the spectral absorptance of the black paint and the spectral transmission of the glass. The overall effect contributes only a small percentage error to the measurements. Each sensor possesses a high level of stability, with the deterioration of the cells resulting in approximately ±1% change in the full scale measurement per year. Finally, the non-linearity of the sensors is a concern, especially with photometers. It is a function of illuminance or irradiance levels. It however tends to contribute only a small percentage error towards the measured values. Table 1.10.2 provides details of the above mentioned uncertainties.

In addition to the above sources of equipment related errors, care must be taken to avoid operational errors such as incorrect sensor levelling and orientation of the vertical sensors, as well as improper screening of the vertical sensors from ground-reflected radiation.

A survey of radiation instruments undertaken by Lof et al. (1965) showed that of the 219 sensors in use across Europe, 65 were CM 11 pyranometers while 107 were the simpler and less expensive Robitzch actinographs with a bimetallic temperature element. The latter instrument is also quite popular in the developing Asian (89 such sensors were reported to be in use), African (16 sensors) and South American (47 sensors) countries where maintenance is often the key factor. The author has visited a solar radiation measurement station in the middle of the Sahara desert and seen the Robitzch actinograph faithfully recording a regular trace of irradiation. The weekly changeover of the recording chart makes this instrument an ideal choice for remote locations. Although not in use with the North American meteorological network, it is known to be used over there in biology and agriculture related work (Coulson, 1975).

Drummond (1965) estimates that accuracies of 2–3% are attainable for daily summations of radiation for pyranometers of first class classification. Individual hourly summations even with carefully calibrated equipment may be in excess of 5%. Coulson (1975) infers that the errors associated with routine observations may be well in excess of 10%. Isolated cases of equipment that is poorly maintained but in the regular network may exhibit monthly-averaged errors of 10% or more. The Robitzch actinograph, even with all the modifications to improve its accuracy, is suitable only for daily summations. At this interval it provides an accuracy of around 10%. However, not all designs of this sensor can claim even this level of accuracy. These figures must be borne in mind when evaluating the accuracy of the relevant computational models.

1.11 Statistical evaluation of models

Checking on the adequacy of the model is important not only in the final stages of the work programme, but more particularly in the initial phase. An examination of residuals is recommended. The procedure is to produce a graph of the residuals d (the difference between observed Y_o and calculated Y_c values of the dependent variable) plotted against the independent variable X or the observed value Y_o.

If the residuals fall in a horizontal band as shown in Figure 1.11.1(a), the model may be judged as adequate. If the band widens as X or Y_o increases, displayed in Figure 1.11.1(b), this indicates a lack of constant variance of the residuals. The corrective

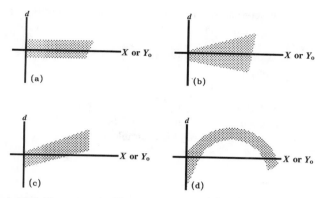

Figure 1.11.1 Plot of residuals for evaluating the adequacy of the model: (a) Adequacy (b) Y_0 needs transformation (c) Missing linear independent variable (d) Missing linear or quadratic independent variable

measure in this case is a transformation of the Y variable. A plot of the residuals such as Figure 1.11.1(c) indicates the absence of an independent variable in the model under examination. If, however, a plot such as Figure 1.11.1(d) is obtained, a linear or quadratic term would have to be added.

In the forthcoming chapters a number of models will be presented wherein one dependent variable is regressed against one or several independent variables. Often correlation between two quantities is also to be examined. In solar energy literature it has become common practice to refer to statistically obtained regression models as 'correlation equations'. Strictly speaking this is wrong usage of statisticians' language. Correlation is the degree of relationship between variables and one seeks to determine how well a linear or other model describes the relationship. On the other hand, regression is a technique of fitting linear or non-linear models between a dependent and a set of independent variables. Thus fitting an equation of the form

$$Y = a_0 + a_1 X \tag{1.11.1}$$

for n pairs of (X, Y) is an example of linear regression. On the other hand, fitting

$$Y = a_0 \exp(a_1 X) + b_0 \sin(b_1 X) \tag{1.11.2}$$

is an example of a non-linear model.

A number of low-priced software packages are available which adequately cover the requirements of fitting linear and non-linear models. The popular spreadsheet packages such as Lotus 1-2-3 (a product of Lotus Corporation) and Excel (a product of Microsoft Corporation) as well as more specialist statistical packages such as SOLO and BMDP (products of BMDP Statistical Software Inc.) are a few examples. For handling very large data arrays one has to resort to FORTRAN and C environments. The text *Numerical Recipes* by Press et al. (1992), with its companion electronic suite of programs, offers solutions at this end. All of the above packages use robust and efficient routines which obviate any particular need for developing optimisation programs from scratch. In the following paragraphs, a brief discussion on the statistical examination of models is provided.

1.11.1 Coefficient of determination r^2

The ratio of explained variation $\Sigma(Y_c - Y_m)^2$ to total variation $\Sigma(Y_o - Y_m)^2$ is called the coefficient of determination r^2. Y_m is the mean of the observed Y values. The ratio lies between zero and one. A high value of r^2 is desirable as this shows a lower unexplained variation.

1.11.2 Coefficient of correlation r

The square root of the coefficient of determination is defined as the coefficient of correlation r. It is a measure of the relationship between variables based on a scale ranging between +1 and −1. Whether r is positive or negative depends on the interrelationship between x and y, i.e. whether they are directly proportional (y increases and x increases) or vice versa.

Once r has been estimated for any fitted model its numerical value may be interpreted as follows. Let us assume that for a given regression model $r = 0.9$. This means $r^2 = 0.81$. It may be concluded that 81% of the variation in Y has been explained (removed) by the model under discussion, leaving 19% to be explained by other factors.

1.11.3 Student's t-distribution

Often the modeller is faced with the question as to what quantitative measure is to be used to evaluate the value of r^2 obtained for any given model (Owen and Jones, 1990). Clearly, r^2 would depend on the size of the data population. For example, a lower value of r^2 obtained for a model fitted against a large data base may or may not be better than another model which used a smaller population. In such situations Student's t-test may be used for comparing the above two models. The following example demonstrates the use of this test of significance for r^2.

Example 1.11.1

For a given location a regression model between average clearness index $\bar{K}_T$ and monthly-averaged sunshine fraction n/N gives $r^2 = 0.64$ for 12 pairs of data points. Using Student's t-test, investigate the significance of r.

The test statistic $t = (n - 2)^{0.5} \{r / \sqrt{(1 - r^2)}\}$, where n is the number of data points and $(n - 2)$ is the degrees of freedom (d.f.). Thus

test statistic $t = (12 - 2)^{0.5} \{0.8/\sqrt{(1 - 0.64)}\} = 4.216$

In this example there are 10 degrees of freedom. Thus from Table 1.11.1 the value of

Table 1.11.1 Percentile values for Student's *t*-distribution

d.f.	P = 0.95	0.98	0.99	0.998	0.999
1	12.706	31.821	63.657	318.310	636.620
2	4.303	6.965	9.925	22.327	31.598
3	3.182	4.541	5.841	10.214	12.924
4	2.776	3.747	4.604	7.173	8.610
5	2.571	3.365	4.032	5.893	6.869
6	2.447	3.143	3.707	5.208	5.959
7	2.365	2.998	3.499	4.785	5.408
8	2.306	2.896	3.355	4.501	5.041
9	2.262	2.821	3.250	4.297	4.781
10	2.228	2.764	3.169	4.144	4.587
15	2.131	2.602	2.947	3.733	4.073
20	2.086	2.528	2.845	3.552	3.850
25	2.060	2.485	2.787	3.450	3.725
30	2.042	2.457	2.750	3.385	3.646
40	2.021	2.423	2.704	3.307	3.551
60	2.000	2.390	2.660	3.232	3.460
120	1.980	2.358	2.617	3.160	3.373
200	1.972	2.345	2.601	3.131	3.340
500	1.965	2.334	2.586	3.107	3.310
1000	1.962	2.330	2.581	3.098	3.300
∞	1.960	2.326	2.576	3.090	3.291

$r = 0.8$ is significant at 99.8% but not at 99.9% (note that for d.f. = 10, $t = 4.216$ lies between 4.144 and 4.587, corresponding to columns for 0.998 and 0.999 respectively). In lay terms this means that using the above regression model, K_T may be estimated with 99.8% confidence.

1.11.4 Mean bias error (MBE) and root mean square error (RMSE)

To enable further insight into the performance evaluation of a model, MBEs and RMSEs may be obtained. These are defined as

$$\text{MBE} = \Sigma (Y_c - Y_o) / n \tag{1.11.3}$$

$$\text{RMSE} = [\Sigma (Y_c - Y_o)^2 / n]^{1/2} \tag{1.11.4}$$

The above formulae provide MBE and RMSE with the same physical units as the dependent variable Y. In some instances non-dimensional MBE (NDMBE) and RMSE (NDRMSE) are required. These are obtained as

$$\text{NDMBE} = \Sigma [(Y_c - Y_o) / Y_o] / n \tag{1.11.5}$$

$$\text{NDRMSE} = \{\Sigma [(Y_c - Y_o) / Y_o]^2 / n\}^{1/2} \tag{1.11.6}$$

1.12 Exercises

1.12.1 Calculate the equation of time EOT and solar declination angle DEC for 12 hours on 15 October 2002 using the FORTRAN executable routines Prog1-3.Exe through to Prog1-6.Exe. Compare your computations with the figures given in Table 1.2.3.

1.12.2 Calculate the solar altitude and solar azimuth for 12 hour local civil time on 15 October 2002 for Belfast (54.583°N, 5.917°W) using Prog1-6.Exe. What would be the error in the estimation of the angles if the difference between the apparent solar time and local civil time was disregarded? Also obtain the incidence angle of the sun's beam on a south-facing vertical surface.

1.12.3 Estimate the twilight illuminance under a clear sky for London (51.5°N, 0.167°W) at 10 p.m. on 21 June. You may use Prog1-6.Exe to obtain the solar altitude (or solar depression angle) and then use Figure 1.8.1 to interpolate the required illuminance.

1.12.4 Long-term records of solar radiation are available for Sutton Bonnington in England (52.833°N, 1.25°W). Calculate the distance between this location and Leicester (52.633°N, 1.083°W) and hence check the validity of the records for the latter site. (Hint: recall that in Section 1.9 the limiting distance for the applicability of use of such data was shown to be 30 km.)

References

Angus, R.C. (1995) *Illuminance Models for the United Kingdom*. PhD thesis, Napier University, Edinburgh.
Astronomical Phenomena (1993) HM Nautical Almanac Office, HMSO, London.
CIBSE (1982) *Guide A2*. Chartered Institution of Building Services Engineers, London.
Coulson, K.L. (1975) *Solar and Terrestrial Radiation*. Academic Press, New York.
Drummond, A.J. (1965) Techniques for the measurement of solar and terrestrial radiation fluxes in plant biological research: a review with special reference to arid zones. *Proc. Montpiller Symp.*, UNESCO.
Duffett-Smith, P. (1988) *Practical Astronomy with your Calculator*, 3rd edn. Cambridge University Press, Cambridge.
Duffie, J.A. and Beckman, W.A. (1980) *Solar Engineering Thermal Processes*. Wiley, New York.
Edgar, S.L. (1992) *FORTRAN for the 90s*. W.H. Freeman, New York.
Foster, N.B. and Foskett, L.W. (1953) A photoelectric sunshine recorder. *Bull. Amer. Met. Soc.* 34, 212.
Hughes, D.W., Yallop, B.D. and Hohenkerk, C.Y. (1989) The equation of time. *Mon. Not. R. Astr. Soc.* 238, 1529.
Kreider, J.F. and Kreith, F. (1981) *Solar Energy Handbook*. McGraw-Hill, New York.

Lamm, L.O. (1981) A new analytic expression for the equation of time. *Solar Energy* 26, 465.

Lof, G.O.G., Duffie, J.A. and Smith, C.O. (1965) World distribution of solar radiation. *Solar Energy* 10, 27.

Nyhoff, L. and Leestma, S. (1995) *FORTRAN 77 and Numerical Methods for Engineers and Scientists*. Prentice-Hall, Englewood Cliffs, N.J.

Observers' Handbook (1969) HMSO, London.

Owen, F. and Jones, R. (1990) *Statistics*. Pitman, London.

Painter, H.E. (1981) The performance of a Campbell-Stokes sunshine recorder compared with a simultaneous record of the normal incidence irradiance. *Meteorological Magazine* 110, 102–87.

Perez, R., Ineichen, P. and Seals, R. (1990) Modelling daylight availability and irradiance components from direct and global irradiance. *Solar Energy* 44, 271.

Press, W.H., Teukolsky, S.A., Vetterling, W.T. and Flannery, B.P. (1992) *Numerical Recipes in FORTRAN: the Art of Scientific Computing*. Cambridge University Press, Cambridge.

Rawlins, F. (1984) The accuracy of estimates of daily global irradiation from sunshine records for the United Kingdom. *Meteorological Magazine* 113, 187.

Rozenberg, G.V. (1966) *Twilight – a Study in Atmospheric Optics*. Plenum, New York.

Tregenza, P.R. (1994) *UK International Daylight Measurement Programme*. Architecture Department, University of Sheffield.

Woolf, H.M. (1968) *Report NASA TM-X-1646*. NASA, Moffet Field, CA.

Yallop, B.D. (1992) *Technical Note*. Royal Greenwich Observatory, Cambridge.

2 DAILY IRRADIATION

Solar energy or daylight utilisation for any site is dependent upon the quantity of the available flux. Obviously, the flux impinging upon any arbitrary surface undergoes monthly as well as diurnal variations. The measurement of the energy received from the sun, on horizontal as well as sloped surfaces, is an expensive affair. As such, few locations in the world have reliable, long-term measured irradiation data sets. Daylight records are even scarcer.

Most radiation data are given as the energy received on a horizontal surface. Since only a very few applications use this configuration, there is a genuine need for insolation estimations to be carried out for sloped surfaces of any given aspect. The accuracy of these models varies from 40–50% for abbreviated techniques to the limits set out by the accuracy of the measuring equipment for modern sophisticated models (Colliver, 1991).

The frequency at which solar radiation data are required depends on the application. While in agricultural meteorology monthly-averaged or even annual energy budget would suffice, detailed simulation studies warrant computation of inclined surface irradiation at an hourly or sub-hourly level. With the increasing interest being shown in photovoltaics (PV), researchers are demanding data to be provided at a minute's frequency. For an abbreviated analysis of solar energy systems, daily values have been used. Yet for other applications monthly-averaged hourly irradiation data are employed.

Measurements by meteorological departments in most countries are made in a manner such that the different climatic and geographical regions are covered. Usually, the number of stations which measure daily global horizontal radiation exceeds the number which report both global and diffuse values. Again, fewer stations measure radiation values on an hourly rather than a day-integrated basis. A typical measurement strategy may be categorised in the following manner:

(a) global horizontal irradiation recorded on a day-integrated basis;
(b) diffuse and global horizontal irradiation recorded on a day-integrated basis;
(c) global horizontal irradiation recorded on an hourly-integrated basis;
(d) diffuse and global horizontal irradiation recorded on an hourly basis;
(e) irradiation recorded on surfaces of several orientations and tilts in addition to normal incidence beam radiation using a pyrheliometer.

The Meteorological Office network for the UK contains 56 stations in the first category (Meteorological Office, 1980a). The number of stations in the second, third, fourth and fifth categories are 23, 26, 19 and 2 respectively (Muneer and Saluja, 1985).

28 SOLAR RADIATION AND DAYLIGHT MODELS

For any locality where a solar energy related simulation study is to be undertaken, the amount of incident solar radiation on a given plane can be predicted with an accuracy dictated by the available data. The prediction will be more refined if it is based on detailed data, e.g. models utilising global and diffuse values will be more accurate than those based on global values alone.

In this chapter those models are presented which enable calculation of diffuse and global horizontal irradiation on a daily, monthly and annual basis.

2.1 Monthly-averaged daily horizontal global irradiation

Initial modelling work carried out in many countries was involved in relating daily horizontal global irradiation to duration of bright sunshine. The first phase of that work involved the development of regression equations from monthly-averaged data. However, work has progressed since then and equations which use data recorded at daily intervals have also been developed. Full exploitation may therefore be made by linking the relationships under discussion with the daily relationship between horizontal diffuse and global irradiation. In the subsequent sections analysis will be presented which enables estimation of diurnal horizontal as well as sloped global and diffuse irradiation. Figure 2.1.1 shows the scheme for estimation of monthly-averaged daily slope irradiation.

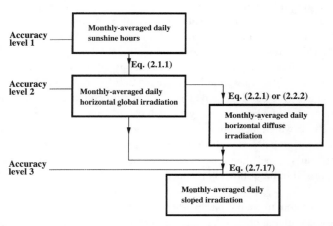

Figure 2.1.1 *Calculation scheme for monthly-averaged daily sloped irradiation*

The original Ångström (1924) regression equation related monthly-averaged daily irradiation to clear day irradiation. However, this method poses the difficulty of defining a clear day. To overcome this difficulty several workers, including Page (1961), Lof et al. (1966), Schulze (1976), Hawas and Muneer (1983), Nagrial and Muneer (1984), Garg and Garg (1985), Turton (1987) and Jain and Jain (1988) have developed relationships of the following form:

$$\bar{G} = \bar{E}\,[a + b(n/N)] \tag{2.1.1}$$

where $\bar{G}$ and $\bar{E}$ are the monthly-averaged daily terrestrial and extraterrestrial irradiation on a horizontal surface, n is the average daily hours of bright sunshine and N is the day length, obtained via

$$\omega_s = \cos^{-1}[-\tan\text{LAT}\,\tan\text{DEC}] \tag{2.1.2}$$

$$N = (2\,\omega_s/15) \tag{2.1.3}$$

Where ω_s, the solar sunrise (or sunset) hour angle, is expressed in degrees. DEC is the solar declination angle obtained via Prog1-3.For, Prog1-5.For or Prog1-6.For (refer to Chapter 1). The ratio n/N is known as the fractional possible sunshine. The extraterrestrial irradiation E (some authors refer to it as the radiation received in the absence of any atmosphere) may be calculated via

$$\begin{aligned}E\,(\text{kW h}/\text{m}^2) = &\,(0.024/\pi)\,I_{SC}\,[1 + 0.033\cos(360\,DN/365)]\\ &\times [\cos\text{LAT}\cos\text{DEC}\sin\omega_s + (2\pi\,\omega_s/360)\sin\text{LAT}\sin\text{DEC}]\end{aligned} \tag{2.1.4}$$

In Eq. (2.1.4) I_{SC} is the solar constant (= 1353 W/m²). Klein (1977) has recommended average days for each month which receive extraterrestrial energy equal to the average energy receipt of the entire month. This is indeed a useful routine since it results in a large time saving. Table 2.1.1 gives the recommended days for each month and provides the values of E for a range of latitudes. These values have been obtained via Prog2-1.Exe. The computational routine is given in Prog2-1.For.

Figure 2.1.2 shows the scatter plot for Eq. (2.1.1) for four Indian locations as presented by Hawas and Muneer (1983). A strong correlation between the two quantities

Table 2.1.1 Monthly-averaged horizontal daily extraterrestrial irradiation (kW h/m²)

	Month and recommended average days per month*											
Lat. (deg)	Jan. 17	Feb. 16	Mar. 16	Apr. 15	May 15	June 11	July 17	Aug. 16	Sep. 15	Oct. 15	Nov. 14	Dec. 10
60	0.942	2.333	4.630	7.548	10.070	11.279	10.687	8.544	5.664	2.995	1.251	0.639
50	2.501	4.012	6.182	8.647	10.579	11.434	11.004	9.404	7.060	4.647	2.851	2.120
40	4.185	5.638	7.547	9.518	10.915	11.483	11.183	10.051	8.243	6.194	4.522	3.791
30	5.850	7.132	8.685	10.119	11.001	11.309	11.125	10.432	9.176	7.574	6.141	5.480
20	7.407	8.436	9.559	10.426	10.806	10.873	10.799	10.524	9.831	8.737	7.631	7.090
10	8.795	9.504	10.144	10.426	10.324	10.171	10.197	10.317	10.188	9.647	8.933	8.550
0	9.961	10.301	10.421	10.119	9.564	9.213	9.330	9.815	10.236	10.272	10.001	9.808

*Recommended average days for each month which receive extraterrestrial energy equal to the average energy receipt for the month (Klein, 1977).

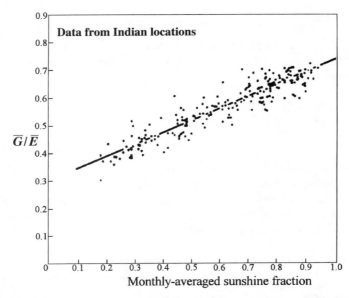

Figure 2.1.2 *Relationship between average clearness index and sunshine fraction*

under discussion is evident. Similar plots, each showing a strong correlation, have been presented by the above researchers.

Whereas the UK Meteorological Office records daily fractional sunshine for 231 locations, the number of stations recording this variable in the United States is around 250. Thus the usefulness of this model lies in its capability to predict daily insolation for all of these sites. Colliver (1991) has presented a most comprehensive monthly-averaged irradiation data set for 1200 locations world-wide using the approach under discussion. Page and Lebens (1986) have also tabulated such data for 13 sites in the United Kingdom.

Table 2.1.2 provides values of the a and b coefficients in Eq. (2.1.1.) for selected sites around the globe, and Prog2-1.For lists the routine for computing $\bar{G}$. The above data for many other sites have been presented by Lof et al. (1966).

The following example demonstrates the use of Prog2-1.Exe.

Example 2.1.1

Using the monthly-averaged sunshine data for June presented in Appendix B, calculate the available horizontal global, diffuse and beam irradiation for London, England. Note that an electronic version of Appendix B is enclosed in the compact disk as File2-1.Csv.

The output generated by Prog2-1.Exe is as follows:

day length = 16.324 h

DAILY IRRADIATION

Table 2.1.2 Coefficients for use in Eq. (2.1.1)

Location	a	b	Source
Stanleyville, Zaire	0.280	0.400	Page (1961)
Nairobi, Kenya	0.240	0.560	
Leopoldville, Zaire	0.210	0.520	
Pretoria, South Africa	0.270	0.460	
Durban, South Africa	0.330	0.350	
Cape Town, South Africa	0.200	0.590	
São Paolo, Brazil	0.240	0.580	
Kew, England	0.150	0.680	
Rothamsted, England	0.160	0.600	
Bracknell, England	0.150	0.700	
Cambridge, England	0.170	0.680	
Garston, England	0.140	0.680	
London, England	0.130	0.650	
Malange, Angola	0.340	0.340	Lof et al. (1966)
Hamburg, Germany	0.220	0.570	
Tamanrasset, Algeria	0.300	0.430	
Buenos Aires, Argentine	0.260	0.500	
Nice, France	0.170	0.630	
Darien, Manchuria	0.360	0.230	
Charleston, SC, USA	0.480	0.090	
Atlanta, GA, USA	0.380	0.260	
Miami, FL, USA	0.420	0.220	
Madison, WI, USA	0.300	0.340	
El Paso, TX, USA	0.540	0.200	
Albuquerque, NM, USA	0.410	0.370	
New Delhi, India	0.341	0.446	Garg and Garg (1985)
Jodhpur, India	0.309	0.484	
Ahmedabad, India	0.302	0.464	
Calcutta, India	0.327	0.400	
Nagpur, India	0.293	0.460	
Bombay, India	0.292	0.464	
Poona, India	0.330	0.453	
Goa, India	0.279	0.514	
Madras, India	0.340	0.399	
Trivandrum, India	0.393	0.357	
Average for 18 Indian locations	0.299	0.448	Hawas and Muneer (1983)
Karachi, Pakistan	0.335	0.391	Nagrial and Muneer (1984)
Lahore, Pakistan	0.326	0.314	
Multan, Pakistan	0.423	0.239	
Quetta, Pakistan	0.436	0.337	
Locations between $20°S$ and $22°N$	0.300	0.400	Turton (1987)
Average for 8 Zambian locations	0.240	0.513	Jain and Jain (1988)
Gebze, Turkey	0.226	0.418	Tiris et al. (1996)

daily extraterrestrial irradiation kW h/m² = 11.413
monthly-averaged global, diffuse, beam irradiation = 4.73, 2.51, 2.21 kW h/m²

The above value for global irradiation is comparable with that given by Cowley (1978), namely 5 kW h/m².

2.2 Monthly-averaged daily horizontal diffuse irradiation

Global irradiation on a slope is the sum of its direct (beam), diffuse and ground-reflected components. Starting from a horizontal global and diffuse irradiation data base, beam irradiation can be evaluated from the difference of the former quantities. This, in turn, can be employed for determining the beam irradiation on a slope. The slope diffuse component is not so straightforward to evaluate. It may be computed from the angular radiance distribution of the sky. The distribution of the sky-diffuse radiance, which is anisotropic in nature, depends on the condition of the sky and determination of it is a fairly involved task. Likewise, the ground-reflected component may be computed given the horizontal diffuse and global irradiation. However, it is also of an anisotropic nature and precise evaluation of it poses difficulties.

It is clear from the preceding discussion that, if the above mentioned approach is adopted, the first step in the estimation of slope irradiation would be to acquire knowledge of horizontal global as well as diffuse irradiation. Therefore in the first instance, for those locations at which only horizontal global irradiation is recorded, a

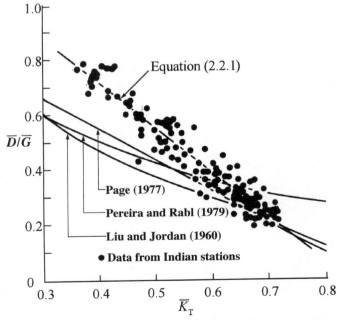

Figure 2.2.1 *Variation of monthly-averaged diffuse ratio against clearness index*

method is needed to estimate the diffuse component. One way is to study the regression between the two quantities at locations where appropriate data are available and hence to establish models which may be used to predict the diffuse component. The models obtained from data recorded on a monthly-averaged basis differ from those obtained from data collected on an individual day-to-day basis.

A regression between monthly-averaged values of diffuse and global irradiation was first developed by Liu and Jordan (1960) in the form of $\bar{D}/\bar{G}$ as a function of clearness index $\bar{K}_T = \bar{G}/\bar{E}$, where $\bar{D}$ is the monthly-averaged daily diffuse radiation incident on a horizontal surface. Many investigators across the world have confirmed the validity of this approach. However, they found that the measured data differ from those predicted by Liu and Jordan's model. As such, questions have been raised about the generality of Liu and Jordan's regression. Figure 2.2.1 shows the regressions proposed by Liu and Jordan (1960) and Pereira and Rabl (1979) for the USA, Page (1961) for the UK and Hawas and Muneer (1984) for India. Page's relationship was derived from shade ring corrected measurements at ten stations. However, no such corrections were applied to the measurements at Blue Hill, Massachusetts which Liu and Jordan used in their work. Pereira and Rabl used beam radiation data measured by pyrheliometer at five stations in the US and hence avoided use of a shade ring correction factor.

Hawas and Muneer's work was based on long-term measurements undertaken at 13 stations in India for the period 1957–75 (Meteorological Office, 1980b). A shade ring correction factor was applied to the diffuse data. Monthly values of the diffuse ratio are shown in Figure 2.2.1 as a function of monthly clearness indices for all months and stations considered. Each point in the figure represents monthly-averaged daily values over the whole period of observation for any given station. The model proposed by Hawas and Muneer for the Indian subcontinent is

$$\bar{D}/\bar{G} = 1.35 - 1.61 \bar{K}_T \tag{2.2.1}$$

Figure 2.2.1 shows that the Indian regression curve differs markedly from the other two regressions, i.e. the diffuse ratio is much higher. Mani and Chacko (1973), in their analysis of solar radiation characteristics for India, also found the diffuse irradiation to be much higher than that for Europe and West Africa. They reported that the highest values of diffuse irradiation occur during the monsoon months when the sky is covered by cloud layers of varying types and densities. The results obtained by Choudhury (1963) for New Delhi also indicate high values of diffuse irradiation.

Pereira and Rabl (1979) have reported a weak effect of changing season upon the regression under discussion. However, no appreciable seasonal variations have been reported by Page (1961) and Hawas and Muneer (1984). This result is in accordance with the study of Stanhill (1966) for Gilat.

To conclude, for the desert and tropical locations which experience high turbidities, Eq. (2.2.1) is recommended. However, for temperate climates and for locations outside the tropics, the following equation given by Page (1977) may be used. This equation is based on eight UK and nine world-wide locations.

$$\bar{D}/\bar{G} = 1.00 - 1.13 \bar{K}_T \tag{2.2.2}$$

Table 2.2.1 Monthly-averaged horizontal daily global and diffuse irradiation (kW h/m²)

	Jan.	Feb.	Mar.	Apr.	May	June	July	Aug.	Sep.	Oct.	Nov.	Dec.
Jersey 49.183° N, 2.183° W												
Global	0.80	1.59	2.86	4.27	5.30	5.96	5.56	4.53	3.36	2.12	0.98	0.64
Diffuse	0.57	0.95	1.46	2.15	2.59	2.68	2.68	2.32	1.68	1.10	0.64	0.44
Easthampstead 51.383° N, 0.783° W												
Global	0.62	1.24	2.32	3.32	4.41	5.22	4.70	3.90	2.92	1.70	0.91	0.53
Diffuse	0.45	0.83	1.36	2.01	2.64	2.76	2.75	2.26	1.69	1.02	0.58	0.38
London 51.517° N, 0.017° W												
Global	0.56	1.14	2.06	3.12	4.10	5.03	4.46	3.58	2.69	1.59	0.81	0.48
Diffuse	0.40	0.73	1.22	1.89	2.40	2.55	2.55	2.07	1.50	0.92	0.53	0.33
Aberporth 52.133° N, 4.567° W												
Global	0.63	1.36	2.58	3.99	4.88	5.54	5.10	4.08	3.07	1.71	0.83	0.53
Diffuse	0.48	0.86	1.43	2.16	2.56	2.76	2.81	2.34	1.69	1.03	0.59	0.39
Cambridge 52.217° N, 0.1° E												
Global	0.61	1.28	2.35	3.28	4.56	5.14	4.65	3.59	2.86	1.66	0.93	0.49
Diffuse	0.45	0.83	1.38	2.00	2.66	2.92	2.76	2.19	1.63	1.00	0.58	0.36
Aldergrove 54.650° N, 6.217° W												
Global	0.48	1.15	2.18	3.63	4.34	5.24	4.37	3.59	2.65	1.38	0.73	0.39
Diffuse	0.33	0.73	1.31	2.08	2.61	2.86	2.76	2.34	1.59	0.89	0.46	0.28
Eskdalemuir 55.317° N, 3.200° W												
Global	0.38	1.12	2.01	3.25	3.90	4.67	4.06	3.42	2.31	1.29	0.65	0.34
Diffuse	0.28	0.68	1.25	1.94	2.40	2.72	2.51	2.16	1.45	0.81	0.41	0.23
Lerwick 60.133° N, 1.183° W												
Global	0.21	0.79	1.76	3.28	3.94	4.81	4.16	3.35	2.06	1.03	0.37	0.14
Diffuse	0.18	0.51	1.11	1.98	2.46	2.72	2.75	2.11	1.35	0.69	0.28	0.12

Example 2.2.1

Using the monthly-averaged measured irradiation data for April presented in Table 2.2.1, calculate the available horizontal diffuse irradiation for Eskdalemuir, Scotland. Compare the result against the corresponding measured value which is also provided in Table 2.2.1.

Prog2-1.Exe estimates $\bar{D} = 1.77$ kW h/m². This may be compared with the measured value of 1.94 kW h/m².

2.3 Annual-averaged diffuse irradiation

Using the analysis presented in Section 2.2 it is logical to investigate an annual regression between the diffuse ratio and the corresponding clearness index. Figure 2.3.1, based on data from widely different locations, indicates a strong correlation between D_{an}/G_{an} and $K_{T,an}$. The following is the regression model:

$$D_{an}/G_{an} = 1 - 1.04\, K_{T,an} \tag{2.3.1}$$

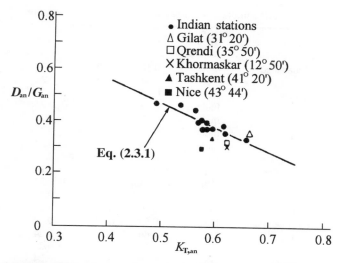

Figure 2.3.1 *Variation of annual-averaged diffuse ratio against clearness index*

Hawas and Muneer (1984) have shown that the ratio of annual global to extraterrestrial irradiation, $K_{T,an}$, varies for the tropics between 0.53 and 0.61. The ratio of annual values of diffuse to extraterrestrial irradiation $K_{D,an}$ varies in a very narrow range, i.e. between 0.22 and 0.25, with an average of 0.233. This value is close to that reported by Stanhill (1966) for Gilat (0.237). The above ratio is also comparable with the values for UK locations obtained by the present author, as may be seen in Table 2.3.1. This is a very interesting phenomenon since the annual-averaged extraterrestrial irradiation may easily be computed (either from Prog2-1.For or Table 2.1.1), and using a universal value of 0.233 for $K_{D,an}$, the annual receipt of diffuse radiation may be obtained for any locality.

Mani and Chacko (1973) have also provided an annual-averaged value of 0.35 for the ratio of annual values of diffuse to global irradiation D_{an}/G_{an}. This is comparable with the values of 0.36 for Gilat (Stanhill, 1966) and 0.3 for Southern Africa (Drummond, 1956).

Example 2.3.1

Using the monthly-averaged data presented in Table 2.2.1, calculate the annual receipt of diffuse irradiation for Easthampstead, England (51.4°N). Use the abbreviated method for computing the annual irradiation quantities shown in Section 2.3.

The abbreviated estimates do not warrant the use of a computer and may easily be carried out on an electronic calculator. Using Table 2.1.1 the annual-averaged daily extraterrestrial irradiation is obtained via interpolation as 6.542 kW h/m². Annual-averaged global irradiation is obtained from Table 2.2.1 as 2.65 kW h/m². Using Eq. (2.3.1), D_{an} is now obtained as 1.53 kW h/m². The measured value of D_{an} from Table 2.2.1 is 1.56 kW h/m².

Table 2.3.1 Annual irradiation data for world-wide locations

Station	Country	$K_{T,an}$	D_{an}/G_{an}	$K_{D,an}$
Ahmadabad	India	0.58	0.38	0.22
Bhaunagar	India	0.61	0.36	0.22
Bombay	India	0.56	0.40	0.22
Calcutta	India	0.53	0.47	0.25
Goa	India	0.59	0.38	0.23
Jodhpur	India	0.65	0.34	0.22
Madras	India	0.57	0.41	0.24
Nagpur	India	0.58	0.38	0.22
New Delhi	India	0.61	0.39	0.24
Poona	India	0.58	0.40	0.23
Shillong	India	0.49	0.47	0.23
Trivandrum	India	0.56	0.45	0.25
Visakhapatnam	India	0.58	0.38	0.22
Gilat	Israel	0.66	0.36	0.24
Jersey	UK	0.47	0.51	0.24
Easthampstead	UK	0.40	0.59	0.24
London	UK	0.38	0.58	0.22
Aberporth	UK	0.44	0.56	0.25
Cambridge	UK	0.41	0.60	0.24
Aldergrove	UK	0.41	0.61	0.25
Eskdalemuir	UK	0.37	0.61	0.23
Lerwick	UK	0.39	0.63	0.24
Average				0.233

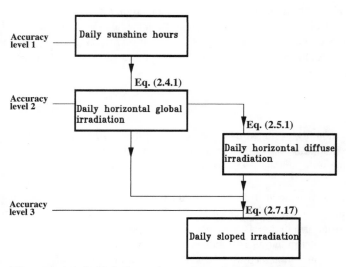

Figure 2.4.1 *Calculation scheme for daily sloped irradiation*

Note that in this $K_{D,an} = 0.238$ which is within 2% of the above quoted universal value of 0.233 (see Table 2.3.1).

2.4 Daily horizontal global irradiation

Figure 2.4.1 shows the scheme for obtaining daily sloped irradiation. As will be demonstrated herein, this scheme uses statistically different models when compared with the models presented in Sections 2.1 and 2.2. A set of linear regression equations between daily global irradiation and the duration of bright sunshine was obtained by Cowley (1978) for ten stations in Great Britain. These equations enable the estimation of daily incident radiation, rather than the monthly-averaged quantity discussed in Section 2.1. Cowley (1978) used the above equations to produce a set of irradiation maps for Great Britain using sunshine data from 132 stations. Cowley's equation is given as

$$G = E\,[d\,\{(a/100) + (b/100)\,(n/N)\} + (1-d)a']\qquad(2.4.1)$$

where $d = 0$ if $n = 0$, otherwise $d = 1$ if $n > 0$, and a' = average ratio of G/E for overcast days. Appendix B provides values of a, and b for use in Eq. (2.4.1) for 15 locations in the United Kingdom. Cowley's choice of Eq. (2.4.1), rather than Eq. (2.1.1), has been justified on the grounds that it results in lower root mean square errors (RMSEs).

Rawlins (1984) has further investigated the accuracy of Cowley's model by its examination against data from 21 independent stations. The reported RMSEs lie in the range 15–20%. If, however, the monthly-averaged daily estimates are to be compared, an accuracy of 3–9% is claimed. An interesting conclusion drawn by Rawlins is that individual daily irradiation is more accurately estimated from local sunshine observations than by assignment from nearby radiometric stations, if these are more than 20 km away. For monthly averages the critical distance increases to about 30 km. It was also shown that the use of sunshine values from a location 50 km away may typically increase the RMSEs of daily estimates from 14% to 22%.

It was shown in Section 2.1 that the regression coefficients for monthly-averaged irradiation are available for world-wide locations (see Table 2.1.2). Cowley's work, however, has not been extended for other geographic regions. At least for the UK, it nevertheless presents an opportunity to compute daily horizontal as well as slope irradiation by linking the above model with those to be presented in Sections 2.5 and 2.7.

Example 2.4.1

On 25 August 1983, 10.1 hours of sunshine were recorded at Easthampstead, England (51.4°N). Calculate the available horizontal global and diffuse irradiation using Cowley's model Eq. (2.4.1).

We note from Appendix B that for a neighbouring site, London, $a = 0.23$, $b = 0.52$ and $a' = 0.14$. Using these coefficients Prog2-2.Exe enables us to obtain horizontal global, diffuse and beam irradiation values of 5.27, 2.21 and 3.06 kW h/m² (daily diffuse irradiation is computed within Prog2-1.Exe using Eq. (2.5.1)). The measured values for daily global and diffuse irradiation were, respectively, 5.37 and 2.06 kW h/m².

2.5 Daily horizontal diffuse irradiation

A regression equation which relates the diffuse fraction of daily global irradiation (diffuse ratio) to the ratio of daily global to extraterrestrial irradiation (clearness index) was also originally developed in the pioneering work of Liu and Jordan (1960). Their correlation is based on data for one location, namely Blue Hill, Massachusetts. No correction was made for the shade ring, and one value of daily extraterrestrial irradiation was used for the middle of the month, for computation of the daily clearness index K_T.

Pereira and Rabl (1979) reinvestigated the correlation of Liu and Jordan using one to four years of pyrheliometric data for five stations in the USA (Albuquerque, New Mexico; Fort Hood, Texas; Livermore, California; Raleigh, North Carolina; and Maynard, Massachusetts). The extraterrestrial irradiation was calculated for each individual day. The latter procedure has been followed by all other investigators who developed similar correlations for different regions. A seasonal trend was reported by Pereira and Rabl. However, the regression equation was fitted for the entire data set as one group.

A parallel study, for the stations mentioned above, was conducted by Erbs et al. (1982) to develop two seasonal correlations, one for winter ($\omega_s < 1.4208$ rad) and the other for the rest of the year ($\omega_s > 1.4208$ rad). Averaged values of the diffuse ratio D/G over finite intervals of K_T were used in the above investigations. Rao et al. (1984) used a year's data for Corvallis, Oregon, to develop an annual as well as a seasonal regression. Individual values of D/G against K_T were used, thus enabling a more representative value for r^2, the coefficient of determination.

Several investigators have developed similar correlations for other geographical regions. The works of Choudhury (1963) and Muneer and Hawas (1984) for India, Ruth and Chant (1976) and Tuller (1976) for Canada, Stanhill (1966) for Israel, Bartoli et al. (1982) for Italy, and Saluja and Muneer (1986) for the United Kingdom are now examined. These studies have confirmed the validity of Liu and Jordan's approach to correlate D/G and K_T.

Choudhury (1963) used only three months of data for New Delhi, and found a higher D/G ratio compared with the Liu-Jordan model. He attributed this to a higher dust content over New Delhi and his small sample of data. Muneer and Hawas (1984) used three years of data for 13 stations in India, spanning from 8.5°N to 28.5°N latitude, to develop correlations for individual stations and for the entire region of India. Individual values of D/G were plotted against K_T and used for regression purposes. The averaged values give a 'false' sense of a high r^2 value for the regressed equation and do not portray a true picture of the scatter of individual points. Muneer and Hawas (1984) found that a good correlation exists between D/G and K_T for the individual locations ($r^2 = 0.893$–0.95) and for the entire Indian region ($r^2 = 0.89$). Figure 2.5.1 shows the regression curves for four Indian locations.

Ruth and Chant (1976) used seven years of data for four Canadian stations (Toronto,

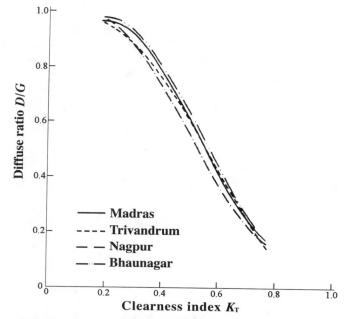

Figure 2.5.1 *Regression curves for daily diffuse ratio, Indian locations*

Montreal, Goose Bay and Resolute Bay). They found that their diffuse ratios were higher than those reported by Liu and Jordan. In a later study Pereira and Rabl (1979) reported the proximity of their results for the US stations to those of Ruth and Chant. A latitude dependence was reported by Ruth and Chant and use of the correlation for individual locations was suggested.

Tuller (1976) made a parallel study for the same four Canadian stations using one year of daily data. He found a latitude effect to be present with the grouped correlation closer to the one due to Choudhury (1963). Tuller also investigated the effect of surface reflectivity on the diffuse irradiation. He concluded that this effect was responsible for only 27% of the variance in diffuse transmission coefficient D/E. Stanhill (1966) studied the relationship using three years of data for Gilat, a semi-arid location in Israel. He also found a departure of the diffuse ratio from that suggested by Liu and Jordan. He attributed his and Choudhury's (1963) results to the higher dust content of stations situated in semi-arid locations.

Bartoli et al. (1982) have studied correlations for Macerata and Genova in Italy using data for six and eight years, respectively. They have reported regressions for these two stations on an individual basis rather than grouping the data together.

Saluja and Muneer (1986) used three years of daily diffuse and global irradiation data from five locations in the United Kingdom (Easthampstead, Aberporth, Aldergrove, Eskdalemuir and Lerwick) to develop individual regression models as well as a single model. The regressed curves for the UK are shown in Figure 2.5.2.

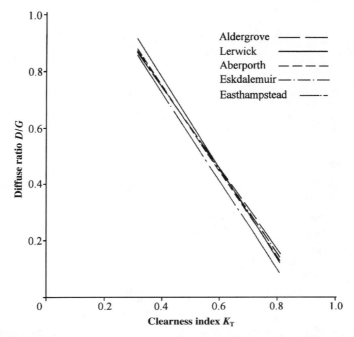

Figure 2.5.2 *Regression curves for daily diffuse ratio, UK locations*

Liu and Jordan used uncorrected diffuse radiation data while other investigators used the corrected data to compensate the obstruction caused by the shade ring or shadow band. A shift of the diffuse ratio curve above that due to Liu and Jordan, first reported by Choudhury (1963) and then by Tuller (1976), can be attributed to this, rather than dust. This observation is made after taking into account the increased diffuse ratios associated with higher turbidities in dust-laden atmospheres. The consequent increment is not so dramatic as to offset the effect of uncorrected diffuse radiation data.

The above paragraphs provide a brief review of the regressions which have been developed between D/G and K_T for various regions in different continents. All of the above mentioned investigators used daily integrated values of diffuse and global solar radiation, rather than hourly or minute-by-minute recordings.

Other investigators have attempted a different approach. Mani and Rangarajan (1983) followed the approach of Hay (1979). Their procedure is as follows. Instead of regressing D/G against K_T, regression is performed for D'/G' against G'/E, where D' and G' are diffuse and global radiation quantities (calculated from sunshine and radiation data) which strike the ground before undergoing multiple reflections with the atmosphere. This is done by using standardised values of atmospheric, cloud base and ground albedo. Hawas and Muneer (1984) in a study for 18 Indian stations found negligible difference between the methods due to Liu and Jordan (1960) and Hay (1979). This may be attributed to the non-existent snow cover at most locations in the tropics.

Smietana et al. (1984), using detailed (1 minute) measurements for Davis, California, have shown that days with disparate weather conditions produced different regressions between D/G and K_T. They have shown that the large scatter around the polynomial regression curve could be reduced by grouping days with equal amounts of percentage sunshine. They concluded in their study that D/G is not a unique function of K_T. However, the use of D/G and K_T ratios on a minute-by-minute basis does not validate their claim in the strict sense as applied to daily values. One may refer to the study of Erbs et al. (1982) in which different regressions were developed for D/G against K_T based on hourly and daily integrated data. A further discussion on the inequality of regressions based on different time scales will be provided in Section 2.6.

Le Baron and Dirmhirn (1983) have used two years of data for two locations in the USA, to investigate the limitations of the Liu–Jordan model for locations with varied snow cover and elevation. The two locations on which their study is based are Salt Lake City (a lower elevation site with relatively short-lived snow cover) and Hidden Peak Site (high elevation with undisturbed snow cover over a large area and time). They report a drastic shift of the Liu–Jordan regression curve under an influence of heavy snow cover. Hay (1979) has attributed this phenomenon to the increased diffuse radiation due to multiple reflections between the snow-covered ground and clouds.

Thus, recapitulating the results of the studies mentioned above, we arrive at the following conclusions:

(a) The validity of Liu and Jordan's approach in considering the correlation between the daily diffuse ratio and the daily clearness index has been confirmed by a large number of investigators in various countries.
(b) The original regression curve proposed by Liu and Jordan was based on uncorrected diffuse irradiation data, and as such, it predicts lower values of the diffuse fraction.
(c) It can be demonstrated via Figure 2.5.3 that without any serious loss of accuracy a global curve may be used for any candidate location.

Figure 2.5.3 shows the regressed curves, discussed above, for Canada, India, the US and the UK. The recommended global model for the diffuse ratio is

$$D/G = \begin{cases} 0.962 + 0.779 K_T - 4.375 K_T^2 + 2.716 K_T^3, & K_T > 0.2 \quad (2.5.1a) \\ 0.98, & K_T < 0.2 \quad (2.5.1b) \end{cases}$$

Example 2.5.1

On 25 August 1983, 5.37 kW h/m² of horizontal global irradiation was recorded at Easthampstead, England (51.4°N). Calculate the diffuse irradiation for this day.

Prog2-2.Exe computes the horizontal diffuse irradiation as 2.17 kW h/m². The corresponding measured value was 2.06 kW h/m². We note by comparing the solutions of Examples 2.4.1 and 2.5.1 that the estimate of diffuse irradiation is significantly improved if we initiate the computation with the measured value of global irradiation

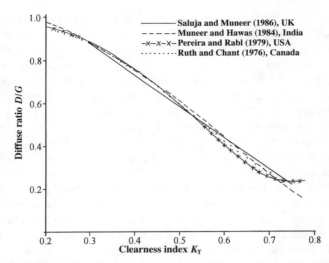

Figure 2.5.3 *Regression curves for daily diffuse ratio, world-wide locations*

rather than sunshine data. This is due to two reasons: (a) the latter computation involves a two-step procedure with the consequent compounding of errors, and (b) the sunshine records are of a much poorer quality than the pyranometric data. Nevertheless, in the absence of any measured irradiation the sunshine data at least enable rough estimates to be made.

2.6 The inequality of the daily and monthly-averaged regressions

As shown above, a number of studies have been carried out in which the daily diffuse component has been regressed against daily global irradiation. These may be broadly classified as:

(a) regressions based on daily values;
(b) regressions based on monthly-averaged values.

A review of the former type of regressions was made in the above sections. In their pioneering work, Liu and Jordan (1960) obtained both of the above mentioned regressions for the United States. Erbs et al. (1982) have re-established these regressions for the United States using pyrheliometric data. Also, Muneer and Hawas (1984) have established these regressions for India.

Monthly-averaged regressions obtained by Liu and Jordan and Erbs et al. were in fact derived from the daily-based regressions in conjunction with the frequency of the daily clearness index. On the other hand Muneer and Hawas have used measured data to obtain both of the above mentioned regressions. The outcome of these studies was that the two relationships differ in their functional form.

Furthermore, for the United Kingdom Page (1977) has developed a monthly-averaged regression while Saluja and Muneer (1986) have developed a daily-based regression. In Section 2.5 this regression was shown to be markedly different from the corresponding daily regression. As pointed out in the preceding section, some authors on occasions have failed to appreciate the fundamental difference between the two relationships under discussion. Muneer and Saluja (1985) have pointed out the erroneous approach of some of the above mentioned studies.

The aim of this section is to provide a theoretical proof, extracted from the works of Muneer (1987) and Saluja et al. (1988), that the relationship between daily quantities differs from that based on monthly-averaged values.

Assume that the relationship between daily diffuse/global irradiation and clearness index is represented by

$$\frac{D}{G} = a_0 + a_1 K_T \tag{2.6.1}$$

where a_0, a_1 are constants, and also that the relationship between monthly-averaged daily diffuse/global irradiation and clearness index is represented by the regression equation

$$\frac{\overline{D}}{\overline{G}} = b_0 + b_1 \overline{K}_T \tag{2.6.2}$$

where b_0 and b_1 are constants.

The relationships for daily and monthly-averaged values would be the same if, and only if, $a_0 = b_0$ and $a_1 = b_1$. The present objective is to prove that the relationships are *not* the same by showing that the assumption $a_0 = b_0$ and $a_1 = b_1$ leads to a contradiction.

If E denotes the daily extraterrestrial irradiation, then

$$\frac{\overline{D}}{\overline{G}} = \frac{\overline{D/E}}{\overline{G/E}} = \frac{\int_0^1 \frac{D}{G} K_T \, df}{\int_0^1 K_T \, df}$$

and also

$$\overline{K}_T = \int_0^1 K_T \, df \tag{2.6.3}$$

In the above integrations, df represents the time increment. For a fuller explanation of the clearness index distribution the reader is referred to Liu and Jordan (1960). Eq. (2.6.2) can be written as

$$\frac{\int_0^1 \frac{D}{G} K_T \, df}{\overline{K}_T} = b_0 + b_1 \overline{K}_T$$

or

$$\int_0^1 \frac{D}{G} K_T \, df = b_0 \overline{K}_T + b_1 \overline{K}_T^2 \qquad (2.6.4)$$

But, using Eq. (2.6.1),

$$\int_0^1 \frac{D}{G} K_T \, df = \int_0^1 (a_0 + a_1 K_T) K_T \, df$$

and so Eq. (2.6.4) becomes

$$\int_0^1 \left(a_0 K_T + a_1 K_T^2\right) df = b_0 \overline{K}_T + b_1 \overline{K}_T^2 \qquad (2.6.5)$$

Now if $a_0 = b_0$ and $a_1 = b_1$, Eq. (2.6.5) becomes

$$\int_0^1 \left(a_0 K_T + a_1 K_T^2\right) df = a_0 \overline{K}_T + a_1 \overline{K}_T^2$$

which reduces to

$$a_1 \int_0^1 K_T^2 \, df = a_1 \overline{K}_T^2$$

or, since $a_1 \neq 0$,

$$\int_0^1 K_T^2 \, df - \overline{K}_T^2 = 0 \qquad (2.6.6)$$

Now the LHS of Eq. (2.6.6) may be proved equal to $\int_0^1 (K_T - \overline{K}_T)^2 \, df$ as follows:

$$\int_0^1 (K_T - \overline{K}_T)^2 \, df = \int_0^1 \left(K_T^2 - 2\overline{K}_T K_T + \overline{K}_T^2\right) df$$

$$= \int_0^1 K_T^2 \, df - 2\overline{K}_T \int_0^1 K_T \, df + \overline{K}_T^2 \int_0^1 df$$

$$= \int_0^1 K_T^2 \, df - 2\overline{K}_T^2 + \overline{K}_T^2$$

Thus

$$\int_0^1 (K_T - \overline{K}_T)^2 \, df = \int_0^1 K_T^2 \, df - \overline{K}_T^2 \qquad (2.6.7)$$

Now $(K_T - \bar{K}_T)^2 > 0$, since $K_T \neq \bar{K}_T$ (see Eq. (2.6.3)). Hence $\int_0^1 (K_T - \bar{K}_T)^2 df > 0$ and thus $\int_0^1 K_T^2 df - \bar{K}_T^2 > 0$ (see Eq. (2.6.7)). But this contradicts Eq. (2.6.6). Therefore, the assumption $a_0 = b_0$ and $a_1 = b_1$ must be false. It follows that the relationships for daily values and monthly-averaged values, represented by Eqs (2.6.1) and (2.6.2) respectively, are different.

The above proof applies for linear relationships between D/G and K_T and between and $\bar{D}/\bar{G}$ and $\bar{K}_T$. Proofs based on polynomial relationships may be obtained by following the above approach. It is worth mentioning that the monthly-averaged relationship between $\bar{D}/\bar{G}$ and $\bar{K}_T$ developed by Page (1977) and the relationship between D/G and K_T established by Saluja and Muneer (1986) in Section 2.5 for the UK are both linear.

2.7 Daily slope irradiation

Daily horizontal global and diffuse irradiation data can be used to estimate slope irradiation provided R, the ratio of global irradiation on a slope to that on a horizontal surface, is known. Mathematically, R is expressed as

$$R = G_{TLT}/G \tag{2.7.1}$$

R can be further decomposed in terms of the contribution of the beam and diffuse components of global irradiation:

$$R_B = B_{TLT}/B \tag{2.7.2}$$

$$R_D = D_{TLT}/D \tag{2.7.3}$$

Thus,

$$R = R_B \, (B/G) + R_D \, (D/G) \tag{2.7.4}$$

The angular correction for the beam component may be solved analytically using the principles of solar geometry. Duffie and Beckman (1991) present equations, based on the work of Liu and Jordan (1962) and Klein (1977), for surfaces sloped towards the equator and also for a sloped surface of any orientation. The former equations are simple but the latter are quite involved, certainly for manual calculations. Muneer and Saluja (1988) have shown that the second set of equations is not applicable for vertical surfaces with north, east and west aspects, i.e. for these aspects indeterminate quantities are generated. Eqs (2.7.5)–(2.7.8), based on the work of Liu and Jordan (1962), Klein (1977) and Muneer and Saluja (1988), enable computation of monthly or daily sloped irradiation on vertical surfaces with eastern, western and northern aspects, and any sloping surface facing south.

For sloped surfaces Liu and Jordan (1962) have shown that

$$R_B = \int_{\omega_s'}^{\omega_s} I_0 \tau \cos \text{INC} \, d\omega / 2 \int_0^{\omega_s} I_0 \tau \sin \text{SOLALT} \, d\omega \tag{2.7.5}$$

Here I_0 is the normal incidence extraterrestrial irradiance, and as it is a constant it cancels out from the numerator and denominator. The atmospheric transmissivity of the beam of irradiance τ is a function of the time of the day and in general R_B can be evaluated only when this functional relationship is known. However, Liu and Jordan (1962) have shown that at the equinox, $\omega_s = \omega_s' = \pi/2$, R_B is independent of τ. Thus for the equinox (21 March and 21 September) τ will cancel out from Eq. (2.7.5). For sloped surfaces in the northern hemisphere facing the equator the expression for R_B may be shown to take the form

$$R_B = \frac{\cos \text{DEC} \cos (\text{LAT} - \text{TLT}) \sin \omega'_s - \omega'_s \sin \text{DEC} \sin (\text{LAT} - \text{TLT})}{\cos \text{LAT} \cos \text{DEC} \sin \omega_s + \omega_s \sin \text{LAT} \sin \text{DEC}} \quad (2.7.6a)$$

and, for surfaces in the southern hemisphere facing the equator,

$$R_B = \frac{\cos \text{DEC} \cos (\text{LAT} + \text{TLT}) \sin \omega'_s - \omega'_s \sin \text{DEC} \sin (\text{LAT} + \text{TLT})}{\cos \text{LAT} \cos \text{DEC} \sin \omega_s + \omega_s \sin \text{LAT} \sin \text{DEC}} \quad (2.7.6b)$$

Liu and Jordan's work is based on an isotropic sky. However, as will be demonstrated in Chapter 4, the isotropic model generates large errors for slope irradiation, in particular on north-facing surfaces. It was shown in Sections 2.4 and 2.5 that the sky-diffuse component is often the major component of the total incident solar energy. As such, Muneer and Saluja (1988) have presented an anisotropic model for obtaining monthly as well as daily slope irradiation. Following the analysis of Muneer and Saluja (1988) the expression for R_B for a vertical surface facing north may be obtained as

$$R_B = \frac{\sin \text{DEC} \cos \text{LAT} \{\omega_s - \cos^{-1}(\tan \text{DEC} / \tan \text{LAT})\} - \cos \text{DEC} \sin \text{LAT} \{\sin \omega_s - \sqrt{(1 - \tan^2 \text{DEC} / \tan^2 \text{LAT})}\}}{\cos \text{LAT} \cos \text{DEC} \sin \omega_s + \omega_s \sin \text{LAT} \sin \text{DEC}} \quad (2.7.7a)$$

For months with little or no receipt of beam radiation a much simpler model has also been proposed:

$$R_B = 0.4 \quad (2.7.7b)$$

Similarly, the expression for R_B for a vertical surface facing east or west is given by

$$R_B = \frac{\cos \text{DEC}(1 - \cos \omega_s)}{2(\cos \text{LAT} \cos \text{DEC} \sin \omega_s + \omega_s \sin \text{LAT} \sin \text{DEC})} \quad (2.7.8)$$

Muneer and Saluja's (1988) model is based on an anisotropic treatment of the sky-diffuse radiation. In the application of this model, fuller details of which are provided in Section 4.3, the fraction of the time any given surface faces the sun (F_{sun}), or remains in shade (F_{shade}), is to be determined. For a vertical surface facing north,

$$F_{sun} = \omega_s - \text{arc} \cos(\tan \text{DEC}/ \tan \text{LAT}) \quad (2.7.9)$$

For any sloped surface facing south,

$$F_{sun} = \omega_s'/\omega_s \tag{2.7.10}$$

where ω_s' is obtained from

$$\omega_s' = \min [\omega_s, \arccos(-\tan(LAT - TLT)\tan DEC)] \tag{2.7.11}$$

The sun's motion is symmetric about east- and west-facing surfaces. Hence for either of the two cases,

$$F_{sun} = 0.5 \tag{2.7.12}$$

When F_{sun} has been found, F_{shade} is obtained as

$$F_{shade} = 1 - F_{sun} \tag{2.7.13}$$

The daily anisotropic model may then be described as follows. For overcast days (that is when $G = D$),

$$G_{TLT} = D(F_{shade}T_{shade} + F_{sun}T_{overcast}) \tag{2.7.14}$$

The 'tilt factors' T_{shade} and $T_{overcast}$ are the ratios of the diffuse irradiation on a slope to that on a horizontal surface. The values of these factors depend upon the tilt of the surface and the radiance distribution of the sky for the respective case. For vertical surfaces $T_{shade} = 0.36$ and $T_{overcast} = 0.4$ (Muneer and Saluja, 1988).

For non-overcast days (that is when $G > D$),

$$G_{TLT} = (G-D)R_B + D\{(R_B F + T_{non-overcast})(1-F)F_{sun} + T_{shade}F_{shade}\} \tag{2.7.15}$$

where $T_{non-overcast}$ is the tilt factor for the sunlit surface under non-overcast conditions (for a vertical surface $T_{non-overcast} = 0.63$). F is the modulating function which 'mixes' the diffuse circumsolar and background-sky irradiation and is given by

$$F = \frac{G-D}{E} \tag{2.7.16}$$

If the ground-reflected irradiation is assumed to be isotropic, the total sloped irradiation is obtained from

$$G_{TLT} = BR_B + DR_D + G_{TLT}[(1 - \cos TLT)/2] \tag{2.7.17}$$

Table 2.7.1 and Figures 2.7.1–2.7.4 enable comparison of the respective performance of the above two procedures in obtaining monthly-averaged and individual daily slope irradiation for Easthampstead and Lerwick. On an individual day-to-day basis the

48 SOLAR RADIATION AND DAYLIGHT MODELS

Table 2.7.1 Monthly-averaged daily irradiation for Easthampstead (51.383°N) (kW h/m²)°

Month	G	D	$G_{TLT,N}$	$G_{TLT,E}$	$G_{TLT,S}$	$G_{TLT,W}$
Measured values						
January	0.62	0.45	0.19	0.36	0.82	0.34
February	1.24	0.83	0.33	0.75	1.49	0.69
March	2.32	1.36	0.52	1.29	2.02	1.18
April	3.32	2.01	0.83	1.80	2.15	1.71
May	4.41	2.64	1.18	2.25	2.11	2.12
June	5.22	2.76	1.41	2.69	2.20	2.53
Computed values, Muneer's model						
January	0.62	0.45	0.18	0.42	1.10	0.42
February	1.24	0.83	0.33	0.82	1.73	0.82
March	2.32	1.36	0.54	1.49	2.42	1.49
April	3.32	2.01	0.80	1.95	2.20	1.95
May	4.41	2.64	1.06	2.46	2.14	2.46
June	5.22	2.76	1.10	2.88	2.13	2.88
MBE, %			−7	15	16	24
Computed values, isotropic model						
January	0.62	0.45	0.22	0.42	0.93	0.42
February	1.24	0.83	0.41	0.81	1.45	0.81
March	2.32	1.36	0.68	1.48	2.07	1.48
April	3.32	2.01	—	1.95	2.01	1.95
May	4.41	2.64	—	2.47	2.11	2.47
June	5.22	2.76	—	2.90	2.20	2.90
MBE, %			9	15	0	25

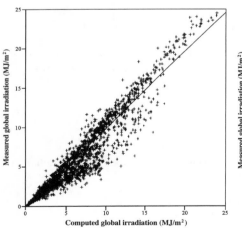

Figure 2.7.1 *Measured versus computed daily sloped irradiation for Easthampstead, UK (isotropic model)*

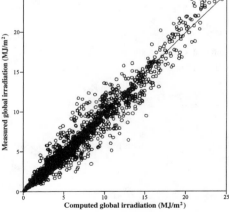

Figure 2.7.2 *Measured versus computed daily sloped irradiation for Easthampstead, UK (anisotropic model)*

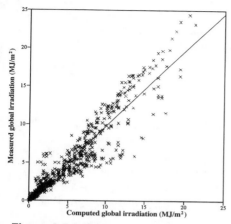 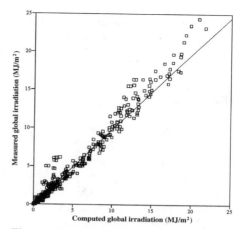

Figure 2.7.3 *Measured versus computed daily sloped irradiation for Lerwick, UK (isotropic model)*

Figure 2.7.4 *Measured versus computed daily sloped irradiation for Lerwick, UK (anisotropic model)*

anisotropic method outperforms the isotropic model. However, as would be expected, the monthly-averaging procedure masks the intricate details. Both of the above procedures are, however, quite restrictive and cumbersome. A much more precise method for obtaining averaged daily sloped irradiation will be introduced in Chapter 4. The procedure requires as a first step the decomposition of averaged daily values into hourly values, to be introduced in Chapter 3.

2.8 Exercises

2.8.1 Monthly-averaged weather data for 15 principal locations are given in Appendix B. Calculate monthly-averaged horizontal daily global and diffuse irradiation for London. Compare your results with the long-term figures reported by the Meteorological Office (1980a), given in Table 2.2.1. You may use Prog2-1.Exe for this task.

2.8.2 Using the computed monthly-averaged horizontal daily global and diffuse irradiation (Exercise 2.8.1) prepare a table for irradiation on vertical surfaces with principal cardinal aspects. You may use Prog2-3.Exe.

2.8.3 On 28 April 1993 the following irradiation values (kW h/m^2) were logged in Edinburgh (55.95°N, 3.2°W):

```
global, horizontal   = 5.883
diffuse, horizontal  = 2.010
north, vertical      = 0.878
east, vertical       = 2.08
south, vertical      = 4.53
west, vertical       = 4.05
```

Use Prog2-4.Exe to estimate vertical slope irradiation for the cardinal aspects. Perform computations using the isotropic and Muneer–Saluja models. Compare your figures against the measured data given above and comment on the error statistics.

References

Ångström, A.K. (1924) On the computation of global radiation from records of sunshine. *Arkiv. for Geof.* 2 (22), 471.
Bartoli, B., Cuomo, V. and Amato, U. (1982) Diffuse and beam components of daily global radiation in Genova and Macerata. *Solar Energy* 28, 307.
Choudhury, N.K.O. (1963) Solar energy at New Delhi. *Solar Energy* 7, 44.
Colliver, D.G. (1991) Solar energy in agriculture. *In Energy in World Agriculture*, ed. B.F. Parker. Elsevier, Amsterdam.
Cowley, J.P. (1978) The distribution over Great Britain of global solar radiation on a horizontal surface. *Meteorological Magazine* 107, 357.
Drummond, A. J. (1956) On the measurement of sky radiation. *Arch. Mat. Wien* B7, 414.
Duffie, J.A. and Beckman, W.A. (1991) *Solar Engineering of Thermal Processes*. Wiley, New York.
Erbs, D.G., Klein, S.A. and Duffie, J.A. (1982) Estimation of the diffuse fraction of hourly, daily and monthly-averaged global radiation. *Solar Energy* 28, 293.
Garg, H.P. and Garg, S.N. (1985) Correlation of monthly-average daily global, diffuse and beam radiation with bright sunshine hours. *En. Conv. & Mgmt* 25, 409.
Hawas, M. and Muneer, T. (1983) Correlation between global radiation and sunshine data for India. *Solar Energy* 30, 289.
Hawas, M. and Muneer, T. (1984) Study of diffuse and global radiation characteristics in India. *En. Conv. & Mgmt* 24, 143.
Hay, J.E. (1979) Calculation of monthly mean solar radiation for horizontal and inclined surfaces. *Solar Energy* 23, 301.
Jain, S. and Jain, P.C. (1988) A comparison of the Ångström-type correlations and the estimation of monthly average daily global irradiation. *Solar Energy* 40, 93.
Klein, S.A. (1977) Calculation of monthly-average insolation on tilted surfaces. *Solar Energy* 19, 325.
Le Baron, B. and Dirmhirn, I. (1983) Strengths and limitations of the Liu and Jordan model to determine diffuse from global irradiance. *Solar Energy* 31, 167.
Liu, B.Y.H. and Jordan, R.C. (1960) The inter-relationship and characteristic distribution of direct, diffuse and total solar radiation. *Solar Energy* 4, 1.
Liu, B.Y.H. and Jordan, R.C. (1962) Daily insolation on surfaces tilted towards the equator. *ASHRAE J.* 3, 53.
Lof, G.O.G., Duffie, J.A. and Smith, C.O. (1966) *World Distribution of Solar Radiation*. Engineering Experiment Station Report 21, University of Wisconsin, Madison, USA.
Mani, A. and Chacko, O. (1973) Solar radiation climate of India. *Solar Energy* 14, 139.
Mani, A. and Rangarajan, S. (1983) Techniques for the precise estimation of hourly values of global, diffuse and direct solar radiation. *Solar Energy* 31, 577.

Meteorological Office (1980a) *Solar Radiation Data for the United Kingdom 1951–75.* MO 912, Meteorological Office, Bracknell.
Meteorological Office (1980b) *Radiation: Short Period Averages 1957–75.* Meteorological Office, Pune, India.
Muneer, T. (1987) *Solar Radiation Modelling for the United Kingdom.* PhD thesis, Council for National Academic Awards, London.
Muneer, T. and Hawas, M. (1984) Correlation between daily diffuse and global radiation for India. *En. Conv. & Mgmt* 24, 151.
Muneer, T. and Saluja, G.S. (1985) A brief review of models for computing solar radiation on inclined surfaces. *En. Conv. & Mgmt* 25, 443.
Muneer, T. and Saluja, G.S. (1988) Estimation of daily inclined surface solar irradiation – an anisotropic model. *Proc. Inst. Mech. Engrs.* 202, 333.
Nagrial, M. and Muneer, T. (1984) Relationship between global radiation and sunshine hours for Pakistan. *Proc. Int. Conf. On Science – Past, Present and Future,* Islamabad, Pakistan.
Page, J.K. (1961) *Proc. UN Conf. on New Sources of Energy*, paper 35/5/98.
Page, J.K. (1977) *The Estimation of Monthly Mean Value of Daily Short Wave Irradiation on Vertical and Inclined Surfaces from Sunshine Records for Latitudes 60°N to 40°S.* BS32, Dept. of Building Science, University of Sheffield.
Page, J.K. and Lebens, R. (1986) *Climate in the United Kingdom.* HMSO, London.
Pereira, M.C. and Rabl, A. (1979) The average distribution of solar radiation – correlations between diffuse and hemispherical and between daily and hourly insolation values. *Solar Energy* 22, 155.
Rao, C.R.N., Bradley, W.A. and Lee, T.Y. (1984) The diffuse component of the daily global solar irradiation at Corvallis, Oregon. *Solar Energy* 32, 637.
Rawlins, F. (1984) The accuracy of estimates of daily global irradiation from sunshine records for the United Kingdom. *Meteorological Magazine* 113, 187.
Ruth, D.W. and Chant, R.E. (1976) The relationship of diffuse radiation to total radiation in Canada. *Solar Energy* 18, 153.
Saluja, G.S. and Muneer, T. (1986) Correlation between daily diffuse and global irradiation for the UK. *BSER&T* 6, 103.
Saluja, G.S., Muneer, T. and Smith, M.E. (1988) Methods for estimating solar radiation on a horizontal surface. *Ambient Energy* 9, 59.
Schulze, R.E. (1976) Physically based method of estimating solar radiation from suncards. *Agricultural Meteorology* 16, 85.
Smietana, P.J., Flocchini, R.G., Kennedy, R.L. and Hatfield, H.L. (1984) A new look at the correlation of K_d and K_t ratio models using one-minute measurements. *Solar Energy* 32, 99.
Stanhill, G. (1966) Diffuse sky and cloud radiation in Israel. *Solar Energy* 10, 96.
Tiris, M., Tiris, C. and Ture, I.E. (1996) Correlations of monthly-average daily global, diffuse and beam radiation with hours of bright sunshine in Gebze, Turkey. *En. Conv. & Mgmt* 37, 1417.
Tuller, S.E. (1976) The relationship between diffuse, total and extraterrestrial solar radiation. *Solar Energy* 18, 259.
Turton, S.M. (1987) Relationship between total radiation and sunshine duration in the humid tropics. *Solar Energy* 38, 353.

3 HOURLY HORIZONTAL IRRADIATION AND ILLUMINANCE

It was pointed out in the previous chapter that irradiation data are usually available as the amount of short-wave energy received on a horizontal surface. The frequency at which solar radiation data are required depends on the application. With the advent of cheap and yet powerful desk-top computers it is now possible to perform energy-system simulations using hourly or sub-hourly data. Such simulations, however, require reliable estimates of slope surface irradiation and illuminance which may be computed from the corresponding horizontal global and diffuse energy data. The UK radiation measurement network is one of the best in Europe, yet long-term hourly data for the latter two quantities are available for only a dozen sites. Therefore, methods are required for estimations to be carried out from long-term records of daily irradiation or other meteorological parameters such as humidity, pressure and sunshine. In this chapter models are presented which enable computation of hourly or instantaneous irradiance and illuminance.

3.1 Monthly-averaged hourly horizontal global irradiation

It has been pointed out earlier that hourly irradiation data lead to more accurate modelling of solar energy processes. Since the daily irradiation is measured more frequently than the corresponding hourly values, it is logical to consider the correlation between the two. Several meteorological stations publish their data in terms of monthly-averaged values of daily global irradiation. Where such measurements are not available, it may be possible to obtain them from the long-term sunshine data via models presented in Chapter 2.

In this section, the correlation between hourly and daily global radiation is examined. The models presented in this and the following section should be used for long-term averaged hourly computations. Figure 3.1.1 shows the calculation scheme for averaged hourly irradiation.

The original work in this field is attributed to Whillier (1956). Liu and Jordan (1960) extended Whillier's work and developed a set of regression curves, shown in Figure 3.1.2, which illustrates the effect of the displacement of the hour from solar noon, and the day length, on the ratio of hourly to daily global irradiation r_G. Collares-Pereira and Rabl (1979) reconfirmed the correctness of the plots of Liu and Jordan and, by using a least-squares fit, obtained

$$r_G = \frac{\pi}{24} (a' + b' \cos \omega) \frac{\cos \omega - \cos \omega_s}{\sin \omega_s - \omega_s \cos \omega_s} \qquad (3.1.1)$$

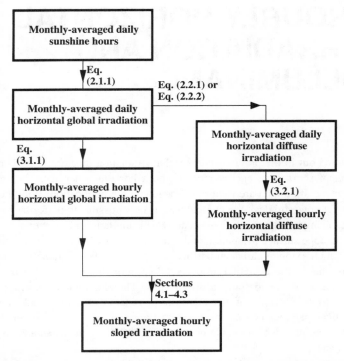

Figure 3.1.1 *Calculation scheme for monthly-averaged hourly sloped irradiation*

where

$$a' = 0.409 + 0.5016 \sin(\omega_s - 1.047) \tag{3.1.2}$$

$$b' = 0.6609 - 0.4767 \sin(\omega_s - 1.047) \tag{3.1.3}$$

Iqbal (1979) tested the applicability of Liu and Jordan's model for three Canadian locations (Toronto, Winnipeg and Vancouver). He found a good agreement, barring the 4.5 h curve. Hawas and Muneer (1984a) used 20 years of averaged data for 13 locations in India to test Liu and Jordan's approach. Figure 3.1.2 shows their data plotted along with Liu and Jordan's regression curves. The scatter of the points is noticeable, and in two cases (2.5 and 3.5 h), a definite shift from Liu and Jordan's regression curves is apparent.

In all the studies reported here, recordings on either side of solar noon have been lumped together. Iqbal (1979) has drawn attention to the problem of asymmetry around solar noon. For Vancouver, he has reported morning values consistently lower than the corresponding afternoon values, while for Montreal no definite pattern was found.

Saluja and Robertson (1983) have also observed that the computed values of yearly long-term averages of irradiation on east- and west-facing surfaces for Aberdeen,

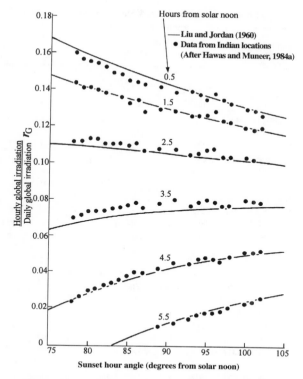

Figure 3.1.2 *Ratio of hourly to daily global irradiation*

Easthampstead and Kew differed from those for other locations in the UK, e.g. higher morning irradiation.

Example 3.1.1

Refer to the extreme right hand column of Table 3.1.1 which provides the monthly-averaged daily global irradiation for Eskdalemuir, Scotland (Meteorological Office, 1980). Evaluate the use of Eq. (3.1.1) to obtain hourly estimates. Note that the time reference in this and the subsequent section is solar time.

Prog3-1.For enables the above estimations to be carried out. An error analysis is also provided. It is evident that barring the sunrise/sunset period the Liu–Jordan model provides reasonable estimates of averaged hourly irradiation for the UK.

3.2 Monthly-averaged hourly horizontal diffuse irradiation

Long-term averages of hourly diffuse irradiation can be computed from monthly-averaged values of daily diffuse irradiation if the ratio of hourly to daily diffuse

56 SOLAR RADIATION AND DAYLIGHT MODELS

Table 3.1.1 Monthly-averaged hourly horizontal irradiation for Eskdalemuir (55.3°N, 3.2°W)

Month		6.5	7.5	8.5	9.5	10.5	11.5	12.5	13.5	14.5	15.5	16.5	17.5	Total $\bar{G}$
Measured Values	(W h/m²)													
January	I_G	0	0	8	36	64	83	81	67	36	8	0	0	381
	I_D	0	0	8	28	47	56	56	47	28	8	0	0	281
February	I_G	3	14	61	125	172	197	194	164	117	58	14	0	1119
	I_D	3	11	42	75	103	117	114	100	72	42	11	0	683
March	I_G	19	81	153	219	267	286	289	258	208	147	75	19	2014
	I_D	17	53	97	133	164	172	169	156	128	94	53	17	1250
April	I_G	92	178	269	333	369	400	403	361	308	244	164	86	3247
	I_D	61	108	156	189	217	233	233	211	181	147	103	61	1936
May	I_G	147	231	297	358	411	422	428	397	358	303	225	144	3900
	I_D	94	142	178	214	242	261	258	244	217	183	139	94	2397
June	I_G	186	281	361	406	458	486	500	464	414	347	269	186	4669
	I_D	114	156	203	236	267	281	281	267	233	197	156	114	2717
Computed Values	(W h/m²)													
January	I_G	0	0	0	35	68	87	87	68	35	0	0	0	380
	I_D	0	0	0	29	50	60	60	50	29	0	0	0	278
February	I_G	0	12	66	121	167	192	192	167	121	66	12	0	1116
	I_D	0	10	47	77	99	110	110	99	77	47	10	0	686
March	I_G	11	73	144	211	265	294	294	265	211	144	73	11	1996
	I_D	10	57	99	133	157	170	170	157	133	99	57	10	1252
April	I_G	88	165	246	319	376	406	406	376	319	246	165	88	3200
	I_D	66	112	153	187	211	223	223	211	187	153	112	66	1904
May	I_G	142	218	295	363	415	442	442	415	363	295	218	142	3750
	I_D	101	144	184	215	238	250	250	238	215	184	144	101	2264
June	I_G	191	270	348	416	467	494	494	467	416	348	270	191	4372
	I_D	123	164	201	231	252	263	263	252	231	201	164	123	2468
Error (%)														
January	I_G	0	0	0	-3	6	4	8	2	-3	0	0	0	0
	I_D	0	0	0	4	6	8	8	6	4	0	0	0	-1
February	I_G	0	-14	8	-3	-3	-3	-1	2	4	13	-14	0	0
	I_D	0	-10	13	3	-4	-6	-3	-1	7	13	-10	0	0
March	I_G	-43	-9	-6	-4	-1	3	2	3	1	-2	-3	-43	-1
	I_D	-40	8	2	0	-4	-1	0	1	4	5	8	-40	0
April	I_G	-4	-7	-9	-4	2	1	1	4	3	1	1	2	-1
	I_D	8	3	-2	-1	-3	-4	-4	0	4	4	9	8	-2
May	I_G	-4	-5	-1	1	1	5	3	4	1	-3	-3	-2	-4
	I_D	7	2	3	1	-2	-4	-3	-3	-1	0	4	7	-6
June	I_G	3	-4	-4	3	2	2	-1	1	1	0	0	3	-6
	I_D	8	5	-1	-2	-6	-6	-6	-6	-1	2	5	8	-9

irradiation r_D is known. Liu and Jordan (1960) have presented a theoretical model,

$$r_D = \frac{\pi}{24} \frac{\cos \omega - \cos \omega_s}{\sin \omega_s - \omega_s \cos \omega_s} \tag{3.2.1}$$

The model shows good agreement with measured data for North America as confirmed by Iqbal (1979).

Collares-Pereira and Rabl (1979) have suggested the possibility of refinement of the above r_D model by incorporating the dependence of the transmission coefficient τ_D on $\bar{K}_T$ and including the angle of incidence in the relationship. In their study of data from 13 Indian locations, Hawas and Muneer (1984a) found considerable data scatter around the r_D model. Indeed, different trends, reflecting a location dependence, were reported by Hawas and Muneer (1984a). Their averaged data points are superimposed on Liu and Jordan's regression curves, shown in Figure 3.2.1. The Indian data points show a compressed range of r_D. Individual monthly values of r_D for the instance of 0.5 solar hour for all Indian locations are shown in Figure 3.2.2. The order of scatter indicates that the r_D model is unsuitable for computing hour-by-hour irradiation.

The effect of $\bar{K}_T$ on r_D was investigated by Hawas and Muneer (1984a). A sample plot is shown in Figure 3.2.3. A consistently decreasing trend of r_D with increasing $\bar{K}_T$ is apparent. The constant value as predicted by the Liu–Jordan model is also shown.

Mani and Rangarajan (1983) have also presented a set of curves for Indian locations, noticeably different from those obtained by Liu and Jordan.

Hay (1976; 1979) attempted to correlate the hourly global and diffuse radiation incident before multiple reflections between the ground and sky, respectively I'_G and I'_D, with the measured hourly global horizontal radiation $\bar{I}_G$ and the fractional possible sunshine $\bar{n}/\bar{N}$. Hay's model is as follows:

$$\bar{I}_G - \bar{I}'_G = \bar{I}_G \rho [\rho_a(\bar{n}/\bar{N}) + \rho_c (1 - \bar{n}/\bar{N})] \quad (3.2.2)$$
$$\bar{I}_D - \bar{I}'_D = \bar{I}_G \rho [\rho_a(\bar{n}/\bar{N}) + \rho_c (1 - \bar{n}/\bar{N})] \quad (3.2.3)$$

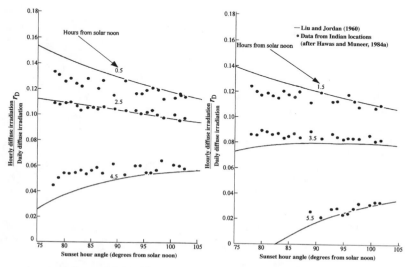

Figure 3.2.1 *Ratio of hourly to daily diffuse irradiation*

58   SOLAR RADIATION AND DAYLIGHT MODELS

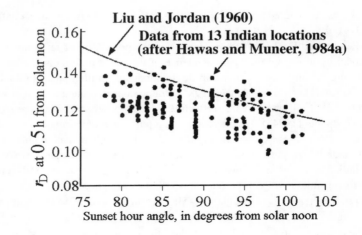

Figure 3.2.2 *Individual (not averaged) values of r_D at 0.5 h from solar noon*

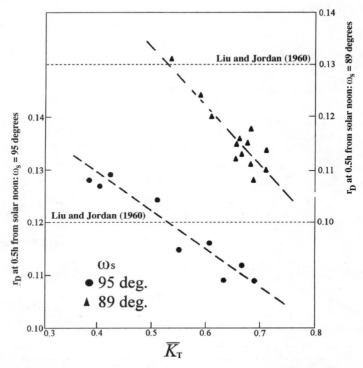

Figure 3.2.3 *Values of r_D at 0.5 h from solar noon for two fixed values of ω_s*

$$\frac{I'_D}{I'_G} = 0.9702 + 1.6688\frac{\bar{I}'_G}{\bar{I}_E} - 21.303\left(\frac{\bar{I}'_G}{\bar{I}_E}\right)^2 + 51.288\left(\frac{\bar{I}'_G}{\bar{I}_E}\right)^3 - 50.081\left(\frac{\bar{I}'_G}{\bar{I}_E}\right)^4 + 17.551\left(\frac{\bar{I}'_G}{\bar{I}_E}\right)^5 \quad (3.2.4)$$

Here ρ, ρ_a and ρ_c are respectively the albedo of the ground, clear sky and clouds. In order to compute the hourly diffuse horizontal radiation $\bar{I}_D$, one has to calculate $\bar{I}'_G$ by Eq. (3.2.2) using $\bar{I}_G$, proceed to compute $\bar{I}'_D$, by Eq. (3.2.4) and then compute $\bar{I}_D$ by Eq. (3.2.3). Iqbal (1983) has compared the accuracy of Liu and Jordan's method with that of Hay's for Canadian locations. The Liu and Jordan model was found to provide better accuracy, in addition to being simpler and straightforward.

Table 3.1.1 enables evaluation of the r_D model of Eq. (3.2.1) for Eskdalemuir data. The estimations were carried out using Prog3-1.For. In this case it may be noted that the accuracy is better than that of the r_G model of Eq. (3.1.1).

Example 3.2.1

Refer to Table 3.1.1. The measured monthly-averaged daily global irradiation $\bar{G}$ for February is given as 1.119 kW h/m². Using Prog2-1.For, with its built-in Eq. (2.2.2), compute the monthly-averaged daily diffuse irradiation $\bar{D}$. Comment on the use of your computed value of $\bar{D}$ to obtain the hourly estimates.

Prog2-1.For provides $\bar{D} = 0.666$ kW h/m². This figure is within 2% of the corresponding measured value. Thus, the hourly estimates may be obtained with almost the same accuracy as those provided in Table 3.1.1.

3.3 Hourly horizontal global irradiation

It was shown above that Liu and Jordan models may only be used to obtain hourly irradiation data from long-term records of monthly-averaged daily values. Building simulation programs however need detailed hour-by-hour data. In the absence of measured irradiation data, a reliable computational method is the meteorological radiation model (MRM). This model estimates the beam transmission through the terrestrial atmosphere based on its attenuation due to mixed gases (such as oxygen, nitrogen and carbon dioxide), water vapour, ozone and aerosols. Figure 3.3.1 shows the schematic for obtaining hourly irradiation from other measured meteorological parameters. The physical basis of the MRM and its performance evaluation is provided in the following sections.

3.3.1 The solar spectrum

Solar spectral wavelengths are measured in micrometres (1 μm = 10^{-6} m), nanometres (1 nm = 10^{-9} m) or ångströms (1 Å = 10^{-10} m). Most broadband solar radiation sensors work in the 300 nm to 3 μm band since this region covers 98% of the energy radiated by the sun. The distribution of the solar spectral irradiance is not uniform in the range 250 nm to 25 μm, i.e. that part of the electromagnetic spectrum which starts with

60 SOLAR RADIATION AND DAYLIGHT MODELS

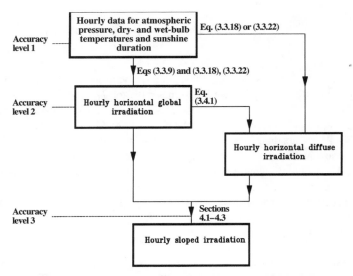

Figure 3.3.1 *Calculation scheme for hourly sloped irradiation*

ultraviolet (UV) radiation and ends in the near infrared (NIR) region. New developments in the measurement of solar spectral distribution are reported by Gueymard (1995). The summation of all energy received at individual wavelengths in the solar spectrum equals the solar constant I_{SC} (=1353 W/m²). This value of I_{SC} has been used throughout this book. This is the irradiance received on a surface normal to the sun's rays at the top of the earth's atmosphere and at a sun–earth distance equal to 1 astronomical unit (1 AU = 1.496×10^{11} m) occurring at the vernal and autumnal equinoxes. The energy of the solar spectrum is approximately distributed as follows: UV 8%, visible band 46% and NIR 46%.

3.3.2 Solar radiation transmission through the earth's atmosphere

To understand the atmospheric transmission of solar radiation it is important to know the composition of the earth's atmosphere. Excluding the water vapour content, the other natural constituents are given in Table 3.3.1 which is based on the *US Standard Atmosphere* (1976). The atmosphere extends to an altitude of 100 km or a pressure of 0.0005 mbar. The naturally occurring gases which play a significant role in the absorption of solar radiation are O_2, N_2 and CO_2 (mixed gases), O_3, H_2O and aerosols.

3.3.2.1 Mixed gases
The depletion of molecular oxygen by solar UV radiation starts at an altitude of 90 km. Therefore, the concentration of O_2 decreases and that of atomic oxygen increases. Because molecular nitrogen is difficult to dissociate, its concentration in the atomic form remains very small even at high altitudes.

3.3.2.2 Ozone
This dry air constituent is created in the upper atmosphere mainly by UV solar radiation.

Table 3.3.1 Normal composition of clean atmosphere

Gas	Content (ppm by volume)
Nitrogen	78 084 000.000
Oxygen	20 948 000.000
Argon	934 000.000
Carbon dioxide	333 000.000
Neon	1 818.000
Ozone	1 200.000
Helium	524.000
Methane	150.000
Krypton	114.000
Hydrogen	50.000
Nitrous oxide	27.000
Carbon monoxide	19.000
Xenon	8.900
Ammonia	0.400
Water vapour	0.400
Sulfur dioxide	0.100
Nitrogen dioxide	0.100
Nitric oxide	0.050
Hydrogen sulfide	0.005

At ground level, it is formed through decomposition of NO (nitrous oxide) that enters the atmosphere as a pollutant. The total amount of ozone (l_o) in the atmosphere as a vertical column is given in units of atmosphere centimetres (atm-cm). This is the height of gaseous O_3 in a vertical column of unit area brought to normal temperature and surface pressure (NTP). Values of l_o are measured world-wide and are readily available, e.g. *Red Book* (1992), or are given by an approximate formula due to Van Heuklon (1979):

$$l_o = J + \{A + C \sin[D(DN+F)] + G \sin[H(LONG+I)]\} [\sin^2(B \text{ LAT})] \quad (3.3.1)$$

where l_o is in milli-atm-cm units, $J = 235$ (the equatorial annual average value of l_o), and A, C, D, F, G, H, I and B are constants given in Table 3.3.2.

3.3.2.3 Water vapour

The amount of water vapour in the entire depth of the atmosphere is referred to as precipitable water l_w. This is the total amount of water vapour (cm) in the zenith direction. l_w can be estimated either by radiosonde data or via any one of the available models. One such model used by Perez et al. (1990), is based on Reitan's (1963) formulation and has been validated for 15 US locations by Wright et al. (1989):

$$l_w = \exp(0.07\text{DPT} - 0.075) \quad (3.3.2)$$

Here DPT is the dew-point temperature, which may be obtained from dry-bulb

Table 3.3.2 Coefficients for use in Eq. (3.3.1)

Coefficient	Northern hemisphere	Southern hemisphere
A	150	100
C	40	30
D	0.9865	0.9865
F	−30	152.625
G	20	20
H	3°	2°
I, longitude > 0	20°	−75°
I, longitude < 0	0°	−75°
J	235	235
B	1.28	1.5
Latitude:	N = +ve, S = −ve	
Longitude:	E = +ve, W = −ve	

temperature and either wet-bulb temperature or relative humidity. The computational scheme and routines are provided in Chapter 7.

3.3.2.4 Aerosols

An aerosol is a small solid or liquid particle that remains suspended in the air and follows the motion of the air stream. Coagulated water vapour molecules following the motion of the air are an example. All suspended particles showing variation in volume, distribution, size, form and material composition are aerosols. They are either of terrestrial origin (e.g. industrial smoke, volcanic eruptions, sandstorms, forest fires) or of marine origin (e.g. salt crystals, ocean spray). Suspended water and ice particles in fog and clouds are also classified as aerosols. In contrast, rain, snow and hail are not aerosol particles. Aerosol particles range in radius from 0.001 to 100 μm.

3.3.2.5 Relative and absolute optical air mass

The attenuation of solar energy by molecules and aerosols in the atmosphere is a function of the type and number of molecules and aerosols in the path of the solar rays. The density of the molecules (or aerosols) multiplied by the path length traversed by a solar ray represents the mass of a substance in a column of unit cross-section. This is called optical air mass or simply air mass. The ratio of the air mass in a slant path in the atmosphere to the air mass in the zenith direction is called the relative air mass m. From the definition it is implied that m takes different values for various solar elevation angles. For a homogeneous atmosphere m may be obtained via Kasten's (1993) formula which provides an accuracy of 99.6% for zenith angles up to 89 degrees:

$$m = [\sin \text{SOLALT} + 0.505\,72\,(\text{SOLALT} + 6.079\,95)^{-1.6364}]^{-1} \quad (3.3.3)$$

This equation is applicable to a standard pressure p_0 of 1013.25 mbar at sea level. For other pressures the air mass may be corrected as

$$m' = m \, (p/1013.25) \tag{3.3.4}$$

where p is the atmospheric pressure (in mbar) at height h (metres above sea level). If p is not known an approximate formula due to Lunde (1980) may be used:

$$p/p_0 = \exp(-0.000\,118\,4\,h) \tag{3.3.5}$$

3.3.3 Attenuation of beam solar radiation

Radiative transfer through terrestrial atmosphere has been discussed at length by Chandrasekhar (1950), Goody (1964), Kondratyev (1969), Barbaro et al. (1979) and several others. When solar radiation enters the earth's atmosphere a part of the incident energy is lost by the mechanisms of scattering and absorption. The scattered radiation is called diffuse radiation, while that part which arrives at the surface of the earth directly from the sun is called direct or beam radiation. According to the Bouguer-Lambert law (*c.* 1760), the attenuation of light through a medium is proportional to the distance traversed in the medium and the local flux of radiation. This law is written as

$$I_B = I_E \exp(-km) \tag{3.3.6}$$

where I_B is the hourly-averaged or instantaneous beam irradiance and I_E is the hourly-averaged or instantaneous extraterrestrial irradiance on a horizontal surface. k is known as the total attenuation coefficient, extinction coefficient or optical thickness. I_E may be computed from

$$I_E = 1353\,[1 + 0.033 \cos(0.017\,202\,4DN)] \sin \text{SOLALT} \tag{3.3.7}$$

Defining the transmission coefficient as $\tau = \exp(-km)$, Eq. (3.3.6) may be written as

$$I_B = I_E \, \tau \tag{3.3.8}$$

3.3.3.1 Rayleigh and Mie scattering

When an electromagnetic wave strikes a particle, part of the incident energy is scattered in all directions. If the particle size is smaller than the wavelength, the phenomenon is called Rayleigh scattering. However, if the particle is of the order of the wavelength the process is known as Mie scattering. Usually air molecules cause Rayleigh and aerosols Mie scattering. Therefore if τ_r and τ_α are the respective transmittances for the Rayleigh and Mie scattering and τ_g, τ_o and τ_w the mixed gases, ozone and water vapour transmittances, Eq. (3.3.8) becomes

$$I_B = I_E \, \tau_r \, \tau_\alpha \, \tau_g \, \tau_o \, \tau_w \tag{3.3.9}$$

3.3.3.2 Ångström's turbidity coefficients for aerosol scattering

Ångström (1929; 1930) suggested a single formula for $k_{a\lambda}$, generally known as Ångström's turbidity formula:

$$k_{a\lambda} = \beta\lambda^{-\alpha} \tag{3.3.10}$$

where β is the Ångström turbidity coefficient, α the wavelength exponent, and λ the wavelength (μm). β is a measure of the amount of the aerosols in the atmosphere in the zenith direction. Typically, it varies in the range 0 to 0.5. α is an index of the size distribution of the aerosol particles. Large values of this coefficient indicate a higher ratio of small to large particles. α typically ranges from 0.5 to 2.5, but a value of 1.3 is commonly employed as originally suggested by Ångström for natural atmospheres. Negative values of this coefficient have however been reported for polluted atmospheres.

The transmittance coefficient due to aerosol scattering may be written as

$$\tau_\alpha = \int_{\lambda=0.3\mu m}^{3\mu m} r_{\alpha\lambda}\, d\lambda = \int_{\lambda=0.3\mu m}^{3\mu m} \exp(-\beta\lambda^{-\alpha} m_\alpha)\, d\lambda \tag{3.3.11}$$

Therefore if estimates of the pair α, β exist for any location, a value of τ_α can be deduced. An alternative way of estimating τ_α is the relationship of Shettle and Fenn (1975), given as

$$k_\alpha = 0.2758\, k_{\alpha, \lambda=0.35\ \mu m} + 0.35\, k_{\alpha, \lambda=0.5\ \mu m} \tag{3.3.12}$$

where $k_{\alpha,\lambda=0.35\ \mu m}$ and $k_{\alpha,\lambda=0.5\ \mu m}$ are the attenuation coefficients due to aerosol scattering at the wavelengths 350 nm (no molecular absorption) and 500 nm (ozone absorption). For the UK these values have been given by Muneer et al. (1996) as 0.72 and 0.56. However, for the *US Standard Atmosphere* these values are, respectively, 0.3538 and 0.2661. It is worth mentioning that geographical variations have only a weak influence on k_α, e.g. its respective values for the UK and the US are 0.394 and 0.387.

3.3.3.3 Atmospheric transmittances

This section provides equations for estimating the transmittances required to obtain the terrestrial beam irradiance. A total of 14 coefficients COF(i), $i = 1,14$, are required in Eqs (3.3.13)–(3.3.17).

According to Iqbal (1983),

$$\tau_\alpha = \exp[-k_\alpha^{\text{COF}(1)} (1 + k_\alpha - k_\alpha^{\text{COF}(2)})\, m'^{\,\text{COF}(3)}] \tag{3.3.13}$$

According to Lacis and Hansen (1974),

$$\tau_o = 1 - [0.1611 x_o(1 + 139.48 x_o)^{-0.3035} - 0.002\,715 x_o(1 + 0.044 x_o + 0.0003 x_o^2)^{-1}],$$
$$x_o = l_o\, m \tag{3.3.14}$$

According to Davies et al. (1975),

$$\tau_r = COF(4) - COF(5)m' + COF(6)m'^2 - COF(7)m'^3 + COF(8)m'^4 \qquad (3.3.15)$$

According to Lacis and Hansen (1974),

$$\tau_w = 1 - COF(9)x_w\,[(1 + COF(10)x_w)^{COF(11)} + COF(12)x_w]^{-1}, \quad x_w = l_w\,m \qquad (3.3.16)$$

According to Bird and Hulstrom (1981),

$$\tau_g = \exp[-COF(13)\,m'^{COF(14)}] \qquad (3.3.17)$$

Numerical values of the 14 coefficients COF(i) for clear, overcast and non-overcast sky conditions are given in Table 3.3.3, for the climates of northern Europe and of southern Europe and the USA.

3.3.4 The meteorological radiation model

The meteorological radiation model (or MRM for brevity) estimates the horizontal beam and diffuse components from just ground-based meteorological data – air temperature, atmospheric pressure, relative humidity (or wet-bulb temperature) and sunshine duration. Such data are readily available world-wide. The MRM is therefore an extremely useful tool. Moreover, the model can estimate the horizontal solar components (diffuse, beam and global irradiance) on an hourly, monthly-averaged hourly, daily or monthly-averaged daily basis. Historically, the development of the MRM may be traced to the works of Chandrasekhar and Elbert (1954), Sekera (1956), Coulson (1959), Dave (1964) and Kambezidis et al. (1996).

Table 3.3.3 Coefficients COF(i) for Eqs (3.3.13)–(3.3.17)

	UK / northern Europe			USA / southern Europe
	Clear	Overcast	Non-overcast	Clear
i	SF = 1	SF = 0	SF > 0	SF = 1
1	1.492 90	0.843 73	2.114 30	0.873
2	0.398 80	0.719 97	0.353 40	0.708 8
3	0.743 83	1.040 70	1.059 30	0.910 8
4	0.892 02	0.049 52	0.832 50	0.972
5	0.074 01	0.068 16	–0.021 60	0.082 62
6	0.009 96	0.010 87	0.017 40	0.009 33
7	0.000 94	0.000 97	0.000 70	0.000 95
8	0.000 18	–0.000 10	0.000 20	0.000 437
9	2.430 00	3.438 15	3.446 20	2.495 9
10	80.187 00	71.764 20	77.024 80	79.034
11	0.578 67	0.131 81	0.414 00	0.682 8
12	6.362 00	5.030 57	3.358 40	6.385
13	0.012 35	0.013 18	0.012 30	0.012 7
14	0.257 81	0.268 15	0.253 80	0.26

3.3.4.1 MRM at an hourly level

This model enables hour-by-hour computation of beam, diffuse and global irradiance. Its accuracy is most precise for clear sky conditions and worst during overcast periods. Some of the air mass and transmission models presented in the above sections have one limitation, i.e. they return unreasonably high values at low solar altitudes. The reader's attention is therefore drawn to the solution incorporated in this text which is to exclude from computations those instances when the solar altitude is less than 7 degrees. This control was found to be quite effective in dealing with broadband as well as spectral irradiance calculations presented in Chapter 5.

MRM for clear skies

The MRM is a broadband empirical algorithm. The clear sky diffuse irradiance model is based on the works of Dave (1979), Bird and Hulstrom (1979) and Pisimanis et al. (1987):

$$I_D = I_E \tau_{a\alpha} \tau_g \tau_o \tau_w \left[\frac{0.5(1-r_r)}{1-m+m^{1.02}} + \frac{0.84(1-r_{as})}{1-m+m^{1.02}} \right] \quad (3.3.18)$$

$$\tau_{a\alpha} = 1 - 0.1(1-\tau_a)(1-m+m^{1.06}) \quad (3.3.19)$$

$$\tau_{as} = 10^{-0.045 m^{0.7}} \quad (3.3.20)$$

The global irradiance on a horizontal surface I_G is given by

$$I_G = (I_B + I_D)\left(\frac{1}{1-r_s r'_\alpha}\right) \quad (3.3.21)$$

where r_s is the ground albedo (a standard value of 0.2 is often quoted) and $r'_\alpha = 0.0685 + 0.17(1-\tau'_\alpha)$ is the albedo of the cloudless sky. Here τ'_α is the Rayleigh scattering transmittance computed at $m = 1.66$.

The performance of the above model is shown in Figure 3.3.2 and Table 3.3.4. It is apparent that under clear skies the model provides extremely satisfactory results. This is to be expected, though, as transmission under such conditions is quite well defined. Contrary to this the overcast skies are the most difficult to model. This is discussed below.

MRM for overcast skies

In this case Eq. (3.3.18) may be used to obtain I_D which also equals the global irradiation I_G. It is well known that overcast skies are most difficult to model owing to their indistinguishable trace on the relevant meteorological variables. A dense, dark overcast and a thin cloud both register nil sunshine duration, but the irradiation would be widely different for these two extremes. Figure 3.3.3 shows the behaviour of the MRM for two UK locations. As would be expected there is considerable scatter in this case. An attempt

Table 3.3.4 Accuracy evaluation of hourly MRM, 1985–94 data

Location	Overcast (0–200 W/m²)			Part overcast (200–600 W/m²)			Clear (600–1000 W/m²)		
	MBE (W/m²)	RMSE (W/m²)	MBE (%)	RMSE (%)	MBE (W/m²)	RMSE (W/m²)	MBE (%)	RMSE (%)	MBE (W/m²)	RMSE (W/m²)	MBE (%)	RMSE (%)

Location	MBE (W/m²)	RMSE (W/m²)	MBE (%)	RMSE (%)	MBE (W/m²)	RMSE (W/m²)	MBE (%)	RMSE (%)	MBE (W/m²)	RMSE (W/m²)	MBE (%)	RMSE (%)
Camborne	33	63	33	63	25	112	6	28	−37	111	−5	14
London	37	61	37	61	52	87	13	22	36	61	4	8
Bracknell	30	57	30	57	30	77	8	19	1	69	0	9
Aberporth	21	46	21	46	4	60	1	15	−51	75	−6	9
Hemsby	32	90	32	90	28	174	7	44	−77	171	−10	21
Finningley	26	54	26	54	26	65	6	16	10	47	1	6
Aldergrove	26	52	26	52	11	65	3	16	−28	69	−4	9
Mylnefield	36	72	36	72	19	115	5	29	−68	122	−8	15
Aberdeen	23	56	23	56	19	69	5	17	−13	69	−2	9
Stornoway	12	38	12	38	6	60	2	15	−33	61	−4	8

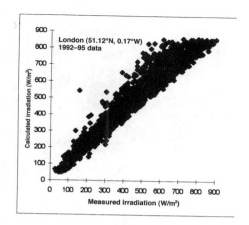

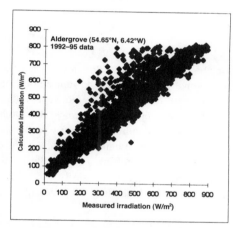

Figure 3.3.2 *Evaluation of the MRM for clear skies*

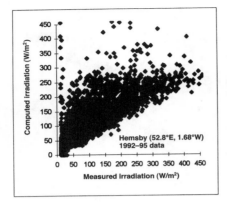 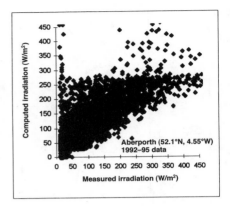

Figure 3.3.3 *Evaluation of the MRM for overcast skies (a) Hemsby (b) Aberporth*

may be made to improve this model via identification of the above two extreme weather conditions. For example, the rain amount may be used as an index to determine the presence of dark overcast. No such attempts have however been reported in the literature. One reason for this may be that the absolute error in energy estimation is low under overcast conditions. It must be borne in mind that the absolute rather than the relative error is the crucial criterion, e.g. the error in calculating daylight illuminance or the energy collection from a solar collector will depend on the absolute deviation of the estimated irradiance. Fortuitously, the MRM offers an almost constant deviation, small relative error for high clear sky irradiance levels and vice versa.

MRM for non-overcast skies

In the above two sections the procedure to obtain irradiation under the two extreme sky conditions was presented. Real skies are, however, rarely clear or completely overcast and hence irradiance models for intermediate skies are very desirable. Some modellers, e.g. Pisimanis et al. (1987), have attempted to obtain intermediate sky irradiation by interpolating the clear and overcast irradiation with the sunshine fraction (SF) as the weighting factor. Muneer et al. (1996) have, however, shown that such an approach leads to excessive scatter of the computed quantities. They have proposed an alternative, more reliable approach which is summarised below.

Muneer et al. (1996) have produced a correlation between hourly diffuse to beam ratio (DBR = I_D / I_B) and beam clearness index ($K_B = I_B / I_E$). Plots for two locations in the United Kingdom are shown in Figure 3.3.4. The generalised correlation is given as

$$\text{DBR} = 0.285 \, K_B^{-1.006} \tag{3.3.22}$$

Figure 3.3.5 shows the results for non-overcast conditions. Table 3.3.4 presents statistical error analysis of the hourly MRM undertaken by Muneer and Gul (1996)

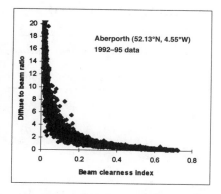
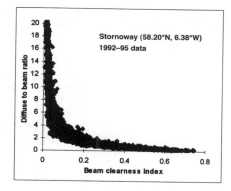

Figure 3.3.4 *Correlation between hourly diffuse and beam irradiation (a) Aberporth (b) Stornoway*

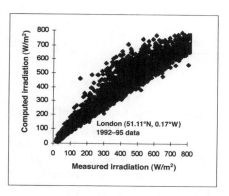
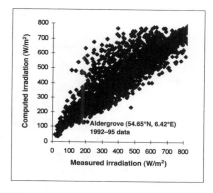

Figure 3.3.5 *Evaluation of the MRM for non-overcast skies (a) London (b) Aldergrove*

against 100 station-year data within the United Kingdom. Broad conclusions which may be drawn from this table are that:

(a) The MBEs suggest that under overcast conditions the model is capable of computing hourly irradiation with an accuracy of 30% (30 W/m^2).
(b) Under part-overcast conditions the errors are in the range of 8% (28 W/m^2).
(c) As would be expected, the model performs best under clear sky regimes, the accuracy in this case being better than 6% (40 W/m^2). The scatter is also kept fairly low for the entire non-overcast set of conditions. This result is in agreement with the works of Gueymard (1993) and Grindley et al. (1995). The latter team have shown that such models can indeed estimate clear sky irradiance to an accuracy of 4%. Their conclusions were based on 1-minute instantaneous measurements undertaken at Cambridge, UK.

Table 3.3.5 Data for Examples 3.3.1 and 3.4.1: London (51.5°N, 0.2°W), 14 April 1995 (Prog3-2.For and Prog3-3.For refer)

Hour	Dry–bulb, temp. (°C)	Wet–bulb temp. (°C)	Atmospheric pressure (mbar)	SF	Measured I_G (W/m²)	Measured I_D (W/m²)	MRM Computed I_G (W/m²)	MRM Computed I_D (W/m²)	Eq. (3.4.1) I_D (W/m²)	Eq. (3.4.2) I_D (W/m²)
6	9.2	7.5	1038	0	66	44	100	100	54	40
7	8.8	7.1	1039	0.4	206	96	203	140	116	78
8	10.0	7.8	1039	1	345	139	435	193	157	123
9	10.7	7.8	1039	0.9	490	159	528	237	170	136
10	12.9	9.3	1039	1	619	170	642	267	167	137
11	14.8	10.1	1038	1	675	195	689	284	185	178
12	15.4	10.7	1037	1	705	204	684	284	183	178
13	16.1	10.7	1037	1	672	203	644	268	187	184
14	16.7	10.8	1036	1	600	198	559	237	183	166
15	17.3	11.2	1035	1	469	199	437	194	189	169
16	17.3	11.4	1034	1	283	184	296	140	190	172
17	17.1	11.5	1033	0	140	122	54	54	121	110
18	16.5	11.3	1033	0	54	51	9	9	49	44
Daily totals				9.3	5324	1964	5280	2406	1951	1715

Example 3.3.1

Table 3.3.5 provides measured hourly weather data for London. Using the MRM, obtain hourly horizontal global and diffuse irradiation.

In cases such as these, Prog3-2.For may be used to generate the output shown in Table 3.3.5. It is evident that for the non-overcast conditions, global irradiation estimates are better than the corresponding diffuse values. This is due to the fact that under such conditions beam irradiation, the dominant component, is obtained first. There is, however, a multiplicative error in the estimation of diffuse irradiation.

3.3.4.2 MRM for daily and monthly-averaged irradiation

The MRM is basically an hourly or a sub-hourly solar radiation model. Its use may also be extended to estimate hourly or sub-hourly illuminance either by fitting the model against concurrent weather and illuminance data, or via the use of luminous efficacy models, to be introduced in Section 3.5. If the model is fitted for illuminance calculations (in this case it would be appropriate to call the model the meteorological illuminance model, MIM) the 14 coefficients provided in Table 3.3.3 would be tuned more closely to the short-wave transmission in the daylight band. Since the visible band represents nearly half of the total energy it would be reasonable to assume that the transmission coefficients would not be too indifferent to those obtained for the broadband model.

The hourly MRM may be used to produce day-integrated irradiation with a high degree of accuracy. Figure 3.3.6 shows a plot which enables such a visual performance evaluation. Tables 3.3.6 and 3.3.7 show the error statistics for monthly-averaged hourly and daily irradiation, obtained via summing the long-term computed and measured values.

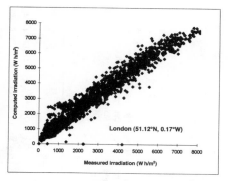 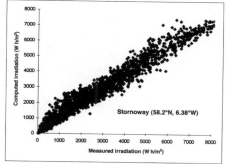

Figure 3.3.6 *Performance of the MRM for daily irradiation*

The MRM is a convenient and powerful tool which enables hourly irradiation and illuminance computations to be carried out for a very large number of locations worldwide, e.g. in the UK alone there are around 300 locations for which the required meteorological data are available. In contrast long-term irradiation data values are available for less than 5% of the above locations.

3.4 Hourly horizontal diffuse irradiation

Sections 3.2 and 3.3 respectively dealt with the estimation of hourly diffuse irradiation from monthly-averaged daily irradiation and meteorological records. The hourly diffuse irradiation on a horizontal surface can be determined more precisely from the records of hourly global irradiation, if validated regression equations relating the two quantities are available. Following the approach of Liu and Jordan (1960) of regressing D/G against K_T, it is logical to correlate hourly diffuse ratio I_D/I_G and hourly clearness index K_t. One such plot for Camborne (UK), based on Muneer's (1987) work, is shown in Figure 3.4.1. The first team to attempt a regression between the above two quantities were Orgill and Hollands (1977). Their study was based on four years of data for Toronto in Canada.

Erbs et al. (1982) followed the procedure of Orgill and Hollands to develop a regression model for the US locations. They used 65 months of data for four locations (Fort Hood, Texas; Maynard, Massachusetts; Raleigh, North Carolina; Livermore, California) with a latitude range of 31–42°N. A single correlation for the four locations showed close agreement with the above mentioned Toronto fit. They also checked the applicability of their regression equation using three years of data obtained for Highett, Australia (latitude 38°S). Erbs et al. proposed that their regression equation was location dependent. In a study for five Australian locations, Spencer (1982) found a similar latitude effect. Spencer compared the performance of models due to Bugler (1977), Boes et al. (1976), Bruno (1978) and Orgill and Hollands (1977). He found the Orgill and Hollands approach to be the best.

Table 3.3.6 Evaluation of the MRM for monthly-averaged hourly irradiation: *measured, †computed for the hour shown, (W h/m²) Bracknell (51.383°N, 0.783°W)

Month	7*	7†	8*	8†	9*	9†	10*	10†	11*	11†	12*	12†	13*	13†	14*	14†	15*	15†	16*	16†	17*	17†
Global irradiation																						
1	0	0	7	0	37	49	78	103	113	141	131	153	125	144	95	111	56	56	17	0	1	0
2	0	0	30	30	89	113	147	170	183	207	200	223	189	218	165	185	119	127	64	56	15	0
3	46	43	104	138	199	225	263	295	302	349	331	370	318	348	277	298	222	232	146	140	67	50
4	136	157	118	265	446	358	377	442	420	493	442	494	424	480	379	436	315	348	236	251	145	139
5	475	227	362	330	421	431	484	508	513	540	523	552	480	526	463	475	404	404	308	318	213	218
6	233	247	582	355	419	433	476	487	501	523	517	550	505	511	475	474	415	417	328	334	238	228
7	216	227	334	328	402	411	472	487	518	536	533	547	526	517	471	479	418	418	338	266	237	0
8	172	183	317	289	382	387	450	461	484	504	511	511	512	496	469	423	359	338	247	184	171	0
9	89	100	282	194	264	284	319	348	392	374	416	356	381	312	443	380	287	287	162	64	80	0
10	34	21	176	113	172	191	235	249	271	279	285	266	237	266	212	243	146	161	70	59	18	0
11	3	0	99	20	74	91	120	139	149	166	170	156	147	147	108	137	50	70	13	0	1	0
12	0	0	30	0	38	46	78	96	106	124	116	132	119	98	74	80	33	17	6	0	0	0
Diffuse irradiation																						
1	0	0	6	0	28	37	55	78	75	100	86	108	82	102	65	82	41	48	14	0	0	0
2	0	0	22	22	63	83	97	122	119	147	132	158	130	153	114	132	83	97	47	50	14	0
3	34	31	79	98	125	147	164	185	183	212	198	221	189	214	168	192	138	154	96	105	49	47
4	85	104	268	160	316	207	219	246	243	270	247	277	270	247	221	247	186	208	144	158	95	99
5	379	145	425	197	207	244	240	262	262	299	270	285	296	273	247	273	208	237	173	191	126	139
6	141	159	192	210	236	255	263	285	281	307	303	315	304	284	270	284	237	251	189	207	141	154
7	125	147	172	199	213	241	248	279	271	290	290	311	285	301	252	280	238	248	183	206	138	153
8	230	117	280	171	326	219	231	256	253	279	258	286	268	280	234	257	210	220	167	174	112	119
9	56	72	129	148	213	175	182	211	202	233	210	242	231	249	178	201	145	164	103	114	56	56
10	23	15	102	78	99	124	132	157	151	176	152	179	167	147	113	140	85	100	47	46	14	0
11	3	0	61	15	51	68	79	101	97	117	101	120	92	107	69	81	38	43	11	0	0	0
12	0	0	7	0	28	34	53	70	70	90	75	95	68	85	51	61	25	16	5	0	0	0

Table 3.3.7 Performance of the MRM for monthly-averaged daily irradiation: m measured, c computed (W h/m²)

Month	Stornoway					Bracknell				
	$\bar{G}_m$	$\bar{G}_c$	$\bar{D}_m$	$\bar{D}_c$	SF	$\bar{G}_m$	$\bar{G}_c$	$\bar{D}_m$	$\bar{D}_c$	SF
1	333	339	338	267	1.0	664	755	452	560	1.8
2	909	934	632	666	2.2	1303	1407	848	977	2.6
3	1906	1976	1189	1320	3.3	2266	2464	1514	1609	3.5
4	3164	3204	2238	2047	4.4	3405	3793	2081	2307	5.0
5	4953	4818	2585	2752	7.1	4638	4823	2917	2829	6.6
6	4577	4505	2786	2946	5.3	4851	4946	3504	3030	6.4
7	3994	3861	2689	2731	4.1	4735	4795	2943	2923	6.8
8	3071	3121	2194	2234	3.6	4170	4172	2612	2497	6.5
9	2365	2331	1514	1570	3.8	2834	2923	1626	1838	4.7
10	1173	1189	783	860	2.5	1706	1802	1244	1183	3.6
11	474	483	344	370	1.4	877	930	700	661	2.3
12	224	195	182	155	0.8	524	577	507	444	1.4

Muneer et al. (1984) used two years of hourly diffuse and global irradiation data from New Delhi to develop an Orgill-Hollands type regression model. The Muneer et al. study provides a link between the works of Orgill and Hollands (1977) and Erbs et al. (1982) for North America and that of Spencer (1982) for Australia. The regression curve for New Delhi was found to lie between the curves for Australia and North America, thus strengthening the assertion of Spencer on location dependency.

Regression curves for world-wide locations are presented in Figure 3.4.2. Muneer (1987) has provided the following equation which was fitted for the mean global curve, and this may be used to estimate the horizontal diffuse irradiance in the absence of any specific regression model for a candidate location:

$$I_D / I_G = 1.006 - 0.317K_t + 3.1241K_t^2 - 12.7616K_t^3 + 9.7166K_t^4 \tag{3.4.1}$$

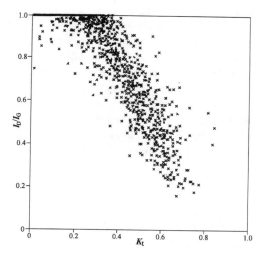

Figure 3.4.1 Hourly diffuse ratio versus clearness index for Camborne, UK

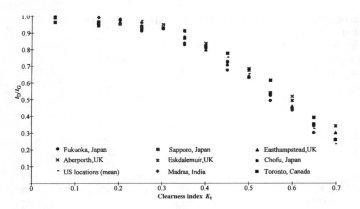

Figure 3.4.2
Hourly diffuse ratio versus clearness index for world-wide locations

Bugler (1977) has followed an alternative approach to obtain the diffuse component from measured global irradiation. Using a year's data for Melbourne, Bugler developed plots of I_D/I_E versus I_G/I_E for a range of solar altitudes. From these, a general set of curves was obtained which showed a linear relationship between I_D/I_E and I_G/I_E. Further, a regression between I_D/I_G and I_G/I_C was obtained, I_C being the hourly clear sky irradiation. Bugler claims that, in this form, the regression equations are independent of location. Although Bugler's method is more involved, its performance has been found to be inferior to that of Orgill and Hollands (1977) as reported by Spencer (1982). Moreover, Bugler's procedure requires monthly data of precipitable water vapour and dust content, a clear disadvantage over the simpler approach of Orgill and Hollands.

The method of Boes et al. (1976) is altogether different from those discussed so far. This method provides regression equations for determining the normal incidence radiation I_0 from the hourly clearness index. The diffuse component is then obtained by subtracting the horizontal component of I_0 from I_G.

Of late there have been some attempts, e.g. Jeter and Balaras (1990) and Perez et al. (1991), to reduce the scatter for the type of regression shown in Figure 3.4.1. The scope of the former study has been rather limited and therefore does not warrant universal applicability. For example, it was shown by Jeter and Balaras (1990) that using the air mass as an additional parameter improves the regression. That study was based on data from one US location. However, other studies, such as the one by Muneer and Saluja (1986) based on data from five UK locations, have specifically shown that solar altitude and hourly sunshine fraction have an insignificant bearing on the relation between the diffuse ratio and the clearness index. Perez et al. (1991) have examined the newer developments in the possible refinements of diffuse (or beam) ratio predictions. Based on data obtained from 14 sites in Europe and the US, they have proposed Maxwell's (1987) model as the overall best. The Maxwell model is given by

$$I_{B,n} = I_0 \{K_{nc} - [A + B \exp (mC)]\} \tag{3.4.2a}$$

$I_{B,n}$ is normal incidence irradiance.

$$K_{nc} = 0.866 - 0.122\, m + 0.0121\, m^2 - 0.000\,653\, m^3 + 0.000\,014\, m^4 \tag{3.4.2b}$$

A, B and C are obtained as follows:

if $K_t \leqslant 0.6$:
$A = 0.512 - 1.560\, K_t + 2.286\, K_t^2 - 2.222\, K_t^3$
$B = 0.370 + 0.962\, K_t$
$C = -0.280 + 0.932\, K_t - 2.048\, K_t^2$

if $K_t > 0.6$:
$A = -5.743 + 21.770\, K_t - 27.490\, K_t^2 + 11.560\, K_t^3$
$B = 41.400 - 118.500\, K_t + 66.050\, K_t^2 + 31.900\, K_t^3$
$C = -47.010 + 184.200\, K_t - 222.000\, K_t^2 + 73.810\, K_t^3$

Reindl et al. (1990) have also examined the influence of other climatic variables to improve the accuracy of the above Orgill-Holland classical approach. They used hourly data from five European and North American locations to develop improved diffuse ratio correlations based on four parameters: K_t, SOLALT, dry-bulb temperature T_{db} and relative humidity RH. Reindl et al. have shown that this approach reduces the residual sum of squares statistic by 14% when compared with the Orgill-Holland type model. In essence Reindl et al.'s work is similar to the MRM model (Section 3.3.4) which estimates the diffuse as well as beam transmission of the short-wave radiation from the latter three of the above parameters. It was shown in Section 3.3.4 that the non-overcast diffuse irradiance may be estimated with an acceptable accuracy from other meteorological data. Thus, Reindl et al.'s model shows promise and may be a candidate for further evaluation. It is as follows:

$$I_D/I_G = 1.000 - 0.232 K_t + 0.0239 \sin \text{SOLALT} - 0.000\,682 T_{db} + 0.0195 \text{RH}, \quad (3.4.3a)$$
$$K_t \leq 0.3, \quad I_D/I_G \leq 1.0$$

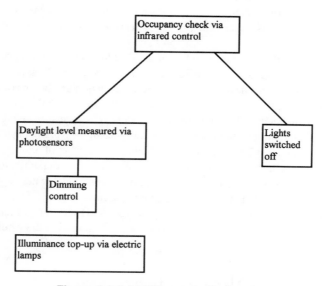

Figure 3.5.1 *Lighting control schematic*

$$I_D/I_G = 1.329 - 1.716K_t + 0.2670 \sin \text{SOLALT} - 0.003\,570T_{db} + 0.1060\text{RH}, \quad (3.4.3b)$$
$$0.3 < K_t < 0.78, \quad 0.1 \leq I_D/I_G \leq 0.97$$

$$I_D/I_G = 0.426K_t - 0.2560 \sin \text{SOLALT} + 0.003\,490T_{db} + 0.0734\text{RH}, \quad (3.4.3c)$$
$$K_t \leq 0.78, \quad I_D/I_G \geq 0.1$$

Example 3.4.1

Hourly global irradiance values were provided for London, UK in Table 3.3.5. Use Eqs (3.4.1) and (3.4.2) to obtain the corresponding diffuse irradiance estimates. Evaluate these estimates against those obtained in Example 3.3.1 using the MRM procedure.

Prog3-3.For may be used to obtain the above estimates for I_D. These estimates are also included in Table 3.3.5. It is evident that using the present diffuse ratio regressions improves I_D estimates. However, given the fact that the MRM does not require any irradiation values as a starting point, its reasonable performance is noteworthy.

3.5 Hourly horizontal illuminance

Recent developments in technology have shown that significant savings in electrical consumption within the building sector are possible through exploitation of daylight. One lighting control schematic which would enable such exploitation is shown in Figure 3.5.1. Modern buildings frequently employ designs and equipment which enable maximum exploitation of solar heat and light with the added possibility of glare and overheat avoidance using weather sensitive, controlled blinds. One such building is the new European Court of Human Rights in Strasbourg which, on demand, can provide up to 92% protection against solar heat (*Building Services and Environmental Engineer*, 1996).

In the United Kingdom the latest update of the Building Regulations aims to improve the lighting efficiency (CIBSE, 1996). To meet the deemed to satisfy approach set out in the new Regulations, the designer is prompted to encourage the maximum use of daylight. Efficient and more precise design of windows in buildings requires the development of inclined surface illuminance models. These in turn require values of horizontal global and diffuse illuminance. In the absence of measurements it is necessary to resort to luminous efficacy models to obtain an estimate of global and diffuse illuminance from other measured or estimated atmospheric parameters. Several approaches have been adopted, including the use of solar altitude, water vapour content, Linke turbidity factor and many other atmospheric parameters.

The global luminous efficacy of daylight K_G is expressed as the ratio of luminance (lx) to irradiance (W/m²) which can be found through the integration of the whole spectrum, i.e.

$$K_G = \left[680 \int_{400\,nm}^{700\,nm} V(\lambda) I_G(\lambda)\,d\lambda \right] \bigg/ \left[\int_0^\infty I_G(\lambda)\,d\lambda \right] \quad (3.5.1)$$

where $V(\lambda)$ is the CIE spectral sensitivity of the human eye and $I_G(\lambda)$ is the solar spectral irradiance.

3.5.1 Atmospheric parameters used in luminous efficacy estimations

Luminous efficacy values enable daylight illuminance to be derived from solar radiation measurements. The problem has been tackled in many different ways by previous researchers. Methods vary by the differing approaches used to account for the absorption and scattering processes that radiation encounters as it passes through the atmosphere. The following paragraph describes the terms and processes occurring in the atmosphere and their effect on luminous efficacy.

In Section 3.3 it was shown that the attenuation of irradiance or daylight is a function of the quantity of atmospheric water vapour and aerosol particles. The presence of dust and soot particles is the result of man-made pollution and natural disasters. A significant contributor is volcanic eruption which may expel millions of tonnes of soot high into the atmosphere where jet streams disperse the particles across the earth.

Particles such as soot and dust are defined as aerosols. The density of dust particles varies with location and season, with higher densities located over land in drier seasons. A turbid atmosphere is renowned for possessing high levels of aerosols resulting in the attenuation or scattering of solar radiation. The scattering process and its effects are discussed in the following sub-sections.

The concept of Ångström's turbidity coefficient β was introduced in Section 3.3.3.2. β may be obtained by assuming an average value of $\alpha = 1.3$ and using visibility measurements vis (expressed in km), obtainable from the local meteorological office via the following equation:

$$\beta = 0.55^{\alpha}(3.912/vis - 0.011\ 62)\ [0.024\ 72(vis - 5) + 1.132] \tag{3.5.2}$$

3.5.1.1 Linke turbidity factor

The Linke turbidity factor T_I allows the estimation of irradiance. However, its use may be extended to estimate illuminance as given by

$$IL_D = I_E \exp(-\alpha_{IL} m T_{IL}) \tag{3.5.3}$$

where IL_D is the horizontal diffuse illuminance.

Navvab et al. (1988) developed an illuminance turbidity coefficient T_{IL} which utilises the Ångström turbidity coefficient β. They have shown that the illuminance turbidity factor may be more sensitive to atmospheric conditions than T_L. T_{IL} is related to the Ångström turbidity coefficient as follows:

$$T_{IL} = 1 + 21.6\ \beta \tag{3.5.4}$$

3.5.1.2 Clouds

The effect of cloud cover for daylight calculations is important. Clouds are good attenuators of solar radiation. It is fair to assume that all the energy removed by the water droplets is in turn scattered.

Tregenza (1980) developed a scheme to estimate illuminance from clouds for use in daylight factors. This method involves, among other considerations, the geometry of clouds. As scattering due to cloud cover will increase the path length of light, the luminous efficacy of an overcast sky, will be slightly higher than that of a clear sky, with values averaging around 110–120 lm / W.

3.5.2 Beam luminous efficacy models

As shown in Section 3.3.3, beam irradiance undergoes strong attenuation as it enters the atmosphere. This process, known as Rayleigh scattering, is prevalent at low solar altitudes. Rayleigh scattering by air molecules tends to be wavelength dependent and in particular affects the visible spectrum. It is estimated that between 10% and 15% of the beam radiation is absorbed as a result of Rayleigh scattering. Another process responsible for the reduction of beam radiation is the effect of Ångström's turbidity coefficient. Again this is most sensitive in the visible waveband Attenuation of beam radiation is also effected by water vapour absorption. This absorption process is mainly confined to the infrared region of the solar spectrum. The overall effect of these processes is the reduction of beam radiation reaching the earth and the reduction of luminous efficacy. Many authors have utilised the physics of these processes in the development of luminous efficacy models and produced good correlations with measured data.

Navvab et al. (1988) developed a semi-empirical formula for a range of turbidities and produced a relationship for the estimation of beam luminous efficacy K_B:

$$K_B = 104.59 [1 - \exp(-9.39 \text{ SOLALT})] \tag{3.5.5}$$

Table 3.5.1 displays the statistical results of all the luminous efficacy models presented herein. Aydinli and Krochmann (1983) developed a polynomial relationship between beam luminous efficacy and solar altitude:

$$\begin{aligned} K_B = {} & 17.72 + 4.4585 \text{ SOLALT} - 8.7563 \times 10^{-2} \text{ SOLALT}^2 \\ & + 7.3948 \times 10^{-4} \text{ SOLALT}^3 - 2.167 \times 10^{-6} \text{ SOLALT}^4 \\ & - 8.4132 \times 10^{-10} \text{ SOLALT}^5 \end{aligned} \tag{3.5.6}$$

A simple yet robust model is the use of a single averaged value of luminous efficacy, a technique that has been adopted by several authors (Littlefair, 1985). An average value of between 93 and 115 1m/W is quoted for beam luminous efficacy. Table 3.5.1 provides the accuracy evaluation of a luminous efficacy value of 104 lm/W.

3.5.3 Global and diffuse luminous efficacy models

Drummond and Ångström (1971) have shown that precise determination of illuminance may be obtained via pyranometric measurements with broadband filters such as the Schott RG8 filter which gives irradiation $I_{G,\,0-700}$ in the $0 < \lambda < 700$ nm interval. This

Table 3.5.1 Comparison of luminous efficacy models against Edinburgh data

	r^2	MBE (lx)	RMSE (lx)
Direct luminous efficacy			
Navvab et al. (1988)	0.973	−258	2339
Aydinli and Krochmann (1983)	0.97	−718	2381
Average (104 lumens / watt)	0.973	−247	2319
Global luminous efficacy			
Aydinli and Krochmann (1983)	0.987	−3016	4351
Chroscicki (1971)	0.985	−5378	6915
Littlefair (1985)	0.987	−767	2842
Average (110 lumens / watt)	0.987	−1667	3277
Diffuse luminous efficacy			
Littlefair (1985)	0.989	−945	1580
Average (120 lumens / watt)	0.986	−82	1870

way all visible wavebands are included but the effects of water vapour in the NIR region are excluded. These authors have proposed a precise equation, with an average error of 1%, to obtain global illuminance:

$$IL_G = 215\, I_{G,\,0-700}\,(1 + 0.102\, m) \qquad (3.5.7)$$

The luminous efficacy of daylight depends upon the way in which the radiant energy is shared between the visible and invisible (infrared and ultraviolet) parts of the spectrum. This in turn depends upon a number of factors including the state of the sky (clear, overcast or average) and the altitude of the sun. In particular, the luminous efficacy is different for the sun alone, the sky alone and for the global radiation (sun plus sky).

Determinations have been made by workers in different parts of the world of the luminous efficacy of daylight from simultaneous measurements of the illuminance and irradiance. Worthy of note among the earliest measurement efforts are Pleijel (1954) in Scandinavia, Blackwell (1954) at Kew, England, Dogniaux (1960) at Uccle, Belgium and Drummond (1958) at Pretoria in South Africa. Pleijel showed that clear and overcast skies vary little in luminous efficacy with solar altitude (and thus with time of year) but that there is a marked decrease in the efficacy of the sun's beam radiation at solar altitudes less than 30°. Blackwell's measurements were related to global radiation with clear, overcast and average skies. The mean efficacies were found to be 119, 120 and 116 lm/W respectively. With average skies, the global efficacy was found to vary between 105 and 128 lm/W.

Moon's (1940) empirical spectral distribution curves for sunlight give an almost constant luminous efficacy of about 117 lm/W for solar altitudes greater than 25°, decreasing to 90 lm/W at 7.5°.

Blackwell does not provide figures for the diffuse luminous efficacy. However, by combining his sky irradiance measurements at Kew with corresponding illuminance measurements reported by McDermott and Gordon-Smith (1951) for neighbouring Teddington, a constant value of 125 lm/W has been deduced by Hopkinson et al. (1966).

Traditionally, researchers have modelled global luminous efficacy for clear and overcast skies. Notable among these are Aydinli and Krochmann (1983) and Chroscicki (1971) who developed formulae relating clear sky global luminous efficacy to solar altitude. However, it is highly desirable to obtain luminous efficacies under all sky conditions. Such models are presented in the following sections.

3.5.4 Littlefair model

The estimation of global luminous efficacy K_G due to Littlefair (1988) involves weighing the sky-diffuse K_D, ground-reflected diffuse K_g and beam luminous efficacies with cloud cover. Diffuse luminous efficacy was reported to be sensitive to cloud cover CC. Littlefair's model is summarised below:

$$K_G = R_B K_B + R_D K_D + R_g K_g \tag{3.5.8}$$

$$K_B = 51.8 + 1.646 \, \text{SOLALT} - 0.015\,13 \, \text{SOLALT}^2 \tag{3.5.9}$$

$$K_D = 144 - 29 \, CC / 8 \tag{3.5.10}$$

where CC is the cloud cover in oktas and may be obtained via its reported relationship with sunshine probability σ,

Table 3.5.2 Coefficients for Perez et al. (1990) luminous efficacy and zenith luminance model (3.5.12)

ε (bin): $i =$	1	2	3	4	5	6	7	8
lower bound	1.000	1.065	1.230	1.500	1.950	2.800	4.500	6.200
upper bound	1.065	1.230	1.500	1.950	2.800	4.500	6.200	—
Global luminous efficacy								
a_i	96.6251	107.5371	98.7277	92.7210	86.7266	88.3516	78.6240	99.6452
b_i	−0.4703	0.7866	0.6972	0.5591	0.9763	1.3891	1.4699	1.8569
c_i	11.5010	1.7899	4.4046	8.3579	7.1033	6.0641	4.9305	−4.4555
d_i	−9.1555	−1.1892	−6.9483	−8.3063	−10.9361	−7.5967	−11.3703	−3.1465
Diffuse luminous efficacy								
a_i	97.2375	107.2129	104.9660	102.3945	100.7100	106.4200	141.8800	152.2300
b_i	−0.4597	1.1508	2.9605	5.5890	5.9400	3.8300	1.9000	0.3500
c_i	11.9962	0.5840	−5.5334	−13.9510	−22.7500	−36.1500	−53.2400	−45.2700
d_i	−8.9149	−3.9490	−8.7793	−13.9052	−23.7400	−28.8300	−14.0300	−7.9800
Zenith luminance								
a_i	40.8646	26.5790	19.3462	13.2425	14.4716	19.7665	28.3923	42.9198
b_i	26.7766	14.7298	2.2895	−1.3987	−5.0932	−3.8843	−9.6634	−19.6247
c_i	−29.5863	58.4662	100.0029	124.7992	160.0932	154.6061	151.5770	130.8072
d_i	−45.7562	−21.2447	0.2547	15.6529	9.1255	−19.2028	−69.3941	−164.0794

$(CC/8) = 1 - 0.55\sigma + 1.22\sigma^2 - 1.68\sigma^3$

K_g in Eq. (3.5.8) is taken as 86 lm/W. R_B, R_D and R_g respectively represent the beam, sky-diffuse and ground-reflected fractions of the global illuminance (or irradiance). A modified version of the above model which does not require sunshine probability was evaluated by Muneer and Angus (1993; 1995), taking instead of Eq. (3.5.10).

$$K_D = (1 - R_D)K_{cl} + R_D K_{oc} \qquad (3.5.11)$$

where K_D is the clear sky luminous efficacy (=144 lm/W) and K_{oc} is the overcast sky luminous efficacy (=115 lm/W). The values of global and diffuse illuminance are then obtained by the multiplication of the appropriate efficacies by the respective irradiance measure. The performance of this modified model was first presented by Muneer and Angus (1993) for Garston data.

Littlefair's (1988) diffuse luminous efficacy model is represented by Eq. (3.5.11). Table 3.5.1 shows the evaluation of this model. Once again good agreement between the estimated and measured data has been found.

3.5.5 Perez et al. model

The Perez et al. (1990) model has a more involved structure and is considered to be more sophisticated. The model is given by

$$K_G \text{ or } K_D = a_i + b_i l_w + c_i \cos z + d_i \ln(\Delta) \qquad (3.5.12)$$

where a_i, b_i, c_i and d_i ($i = 1$ to 8) are the coefficients of the 4 × 8 matrix given in Table 3.5.2. ε represents the sky clearness from overcast sky through partly cloudy conditions to clear skies,

$$\varepsilon = [\{(I_D + I_{Bn})/I_D\} + kz^3] / [1 + kz^3] \qquad (3.5.13)$$

where I_{Bn} is the normal incidence irradiance, z is the solar zenith angle and k is a constant equal to 1.041 for z in radians. Perez et al. (1990) regard the variations in ε as reflecting the changing atmospheric turbidity, and the sky brightness coefficient Δ as denoting the optical transparency of the cloud cover. The equation for sky brightness is

$$\Delta = I_D m/I_E \qquad (3.5.14)$$

The atmospheric precipitable water content l_w is obtained from Eq. (3.3.2).

The Perez et al. model was derived empirically on the basis of data recorded at ten US and three central European sites operating mainly hourly, with some stations recording at 15 minute intervals. The periods of measurements ranged from six months to three years.

Muneer and Angus (1993) have evaluated the above all-sky models (Littlefair, 1988; Perez et al. 1990) using data obtained from Watford (North London). Figure 3.5.2 and

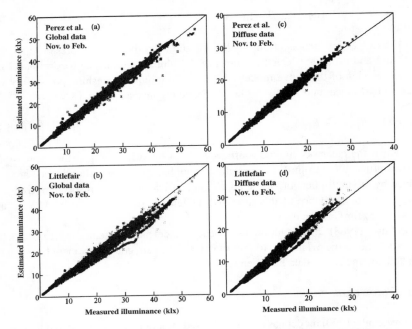

Figure 3.5.2 *Performance of luminous efficacy models*

Table 3.5.3 Performance of Perez et al. (1990) and Littlefair (1988) luminous efficacy models, North London data (lx)

Model	Statistic	Illuminance Global	Diffuse
Perez et al.	MBE	410	180
	RMSE	1160	660
Littlefair	MBE	−1120	100
	RMSE	2210	990

Table 3.5.3 present the measured and modelled horizontal illuminances. The Littlefair model has a tendency to underestimate global illuminances. Some overestimation of the diffuse illuminance is also evident. The data scatter in the two cases is of the same order. The Perez et al. model was reported to perform exceptionally well in both the global and the diffuse illuminance estimations. Moreover the data scatter is minimal. Statistical tests for the above models are given in Table 3.5.1.

Average luminous efficacy models also show promise, at least for the temperate belt of the globe. Muneer and Angus (1995) have shown that for UK locations the respective average luminous efficacies of 110 lm/W and 120 lm/W for global and diffuse components give results which are comparable with most of the diffuse illuminance

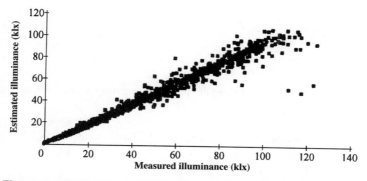

Figure 3.5.3 *Performance of average global luminous efficacy model (luminous efficacy = 110 lm/W)*

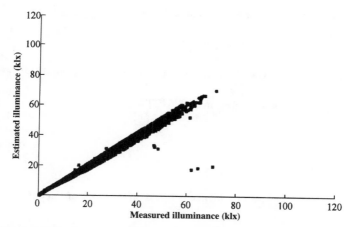

Figure 3.5.4 *Performance of average diffuse luminous efficacy model (luminous efficacy = 120 lm/W)*

models presented herein. In this respect attention is drawn to Figures 3.5.3 and 3.5.4 and Table 3.5.4 (*see p.88*).

3.5.6 DuMortier–Perraudeau–Page model

This model has been used in the preparation of the *European Solar Radiation Atlas* (CEC, 1996) and *CIBSE Guide J* (CIBSE, 1997). The diffuse K_D, beam K_B and global K_G luminous efficacy models are given below.

The luminous efficacy of diffuse irradiance is obtained via the work undertaken by Chauvel (1993). The pseudo cloud cover *CC* is given by

$$CC = 1 - 0.55\, NI + 1.22\, NI^2 - 1.68\, NI^3 \tag{3.5.15}$$

where *NI* is the nebulosity index (DuMortier, 1994a; 1994b). Then for $h > 5°$, $K_D = 144 - 29\ CC$ \hfill (3.5.16)

else, $K_D = 120$ \hfill (3.5.17)

The following steps highlight the computation of the nebulosity index NI. The air mass m is given by

$$m = [\sin \text{SOLALT} + 0.505\ 72 \exp\{-1.6364 \ln(\text{SOLALT} + 6.079\ 95)\}]^{-1}$$

The Rayleigh scattering coefficient, Ar is given by

$$Ar = \{5.4729 + m\ [3.0312 + m\ \{-0.6329 + m\ (0.091 - 0.005\ 12m)\}]\}^{-1}$$

The theoretical horizontal diffuse illuminance for clear sky $I_{D,cl}$ is given by

$$I_{D,cl} = 0.0065 + (0.255 - 0.138 \sin \text{SOLALT}) \sin \text{SOLALT}$$

The theoretical cloud ratio CR for clear sky with a turbidity factor of 4 is given by

$$CR = I_{D,cl} / [I_{D,cl} + \exp(-4m\ Ar) \sin \text{SOLALT}]$$

Finally,

$$NI = [1 - (I_D / I_G)] / (1 - CR)$$

The luminous efficacy of beam irradiance is based on the work of Page (CEC, 1996) as follows:

$$K_B = 62.134 - 0.758\ 85\ \text{SOLALT} + 0.277\ 49\ \text{SOLALT}^2 - 0.012\ 108\ \text{SOLALT}^3$$
$$+ 0.000\ 205\ 2\ \text{SOLALT}^4 - 1.2278 \times 10^{-6}\ \text{SOLALT}^5,$$
$$\text{SOLALT} \leq 50° \hfill (3.5.18)$$

$K_B = 103 + 0.2\ (\text{SOLALT} - 50)$, $50° < \text{SOLALT} \leq 60°$ \hfill (3.5.19)

$K_B = 105$, \hfill $\text{SOLALT} > 60°$ \hfill (3.5.20)

The luminous efficacy of global irradiance is obtained as the weighted average of K_B and K_D:

$$K_G = (K_B \cdot I_B + K_D\ I_D) / I_G \hfill (3.5.21)$$

3.5.7 Other approaches

Delaunay (1995) and Muneer (1995) have shown that for improved precision over an average-value model the global and diffuse efficacies may be regressed against the instantaneous clearness index K_t. It has been reported in the literature that the average values of the above efficacies vary with the geographic location. Whereas Littlefair (1988) cites an average overcast and clear sky diffuse luminous efficacy of 115 lm/W and 144 lm/W respectively, Koga et al. (1993a) report a range of 100–140 lm/W for overcast conditions and 153–165 lm/W for clear skies.

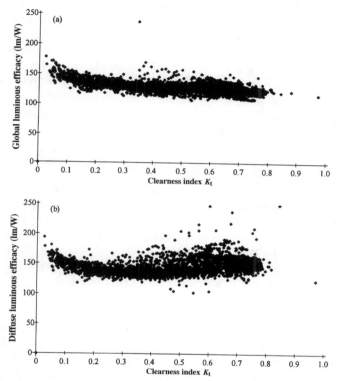

Figure 3.5.5 *Relationship between (a) global and (b) diffuse luminous efficacy and clearness index at Fukuoka*

Figure 3.5.5 shows the variation of global and diffuse efficacies plotted against the clearness index K_t for a Japanese location (Fukuoka, 33.5°N, 130.5°E). Further plots for other Japanese locations have been presented by Muneer (1995). Herein, four points are worth noting:

(a) There is a definite relationship between the dependent quantities (global and diffuse luminous efficacies) and the independent parameter K_t.
(b) The global efficacy shows a decreasing trend (see Figure 3.5.5a). As reported by Perez et al. (1990) the increase in global efficacy for overcast conditions is attributable to the increase in water vapour absorption. For the US and Swiss locations the global efficacy was shown to rise up to 140 lm/W. This trend has also

been noted by Littlefair (1988). On the other hand the efficacy in the absence of any atmosphere is 97 lm/W.
(c) Japanese locations display a higher value of efficacy for the overcast conditions than found elsewhere. This was also the conclusion drawn by Koga et al. (1993a; 1993b).
(d) The diffuse efficacy displays a sag in the middle section of the plot (part-overcast conditions) with higher values at the two extremes of heavy overcast and clear skies. The increase in the diffuse efficacy of clear skies has been explained by the increased contribution of molecular (Rayleigh) scattering. This was confirmed to be the case for the US and Swiss locations (Perez et al. 1990). It was noted by Muneer (1995) that $K_t = 0.4$ was the local threshold for the onset of thin overcast. As explained above, an increase in cloud cover (bright thin clouds to heavy overcast) results in an increase in diffuse luminous efficacy (once the sky is overcast the global and diffuse efficacies are identical). This phenomenon explains the sagging curves shown in Figure 3.5.5b.

It is clear from the above analysis that one single fit for the efficacy models, to cover all geographic areas of the globe, would pose difficulties. The present author has evaluated the validity of the Perez et al. model. Figure 3.5.6 shows its performance for estimating global and diffuse illuminance from their respective irradiance measures. The model consistently underpredicts both global and diffuse illuminance. However, the low scatter indicates that the model would perform adequately if it is fitted against local data. The MBE and RMSE obtained for the Perez et al. model against Fukuoka data were -12% and 13% (Muneer, 1995). A further point of interest noted by Muneer (1996) was that the use of a constant value of $l_w = 2$ cm does not have any appreciable influence on the sensitivity of the Perez et al. model.

In conclusion, all of the above luminous efficacy models, i.e. average value, Littlefair and Perez et al., are good estimators so long as their performance has been evaluated against (at least) short-term data for any given location. They are not, however, universal in their nature and thus may need fitting against the local conditions. For the climates of North America and Europe, the Perez et al. model may be used with confidence. Prog3-4.For provides a routine for the Perez et al. model which enables estimation of global and diffuse illuminance from the corresponding irradiance data.

3.5.8 Zenith luminance

Zenith luminance is an important design parameter and is also required to obtain the absolute luminance distribution of the sky hemisphere. The sky radiance and luminance distributions will be discussed in Chapter 4. A number of researchers have proposed zenith luminance models for clear skies. Notable among these are Kittler (1970), Krochmann (1970), Nagata (1970), Liebelt (1975), Dogniaux (1979), Karayel et al. (1983), Nakamura and Oki (1986) and Rahim et al. (1993). Some of the above models are not applicable in the tropical belt as they use a tangent function of the solar altitude (Rahim et al., 1993). The latter team have also demonstrated the site specific nature of many of the older models.

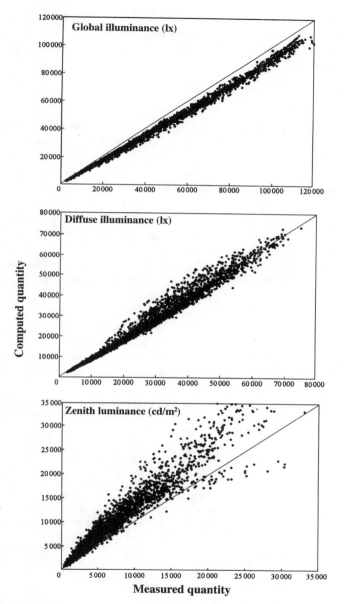

Figure 3.5.6 *Evaluation of Perez et al. (1990) model against Fukuoka data (1994)*

The relative luminance distribution of the CIE standard clear sky (CIE, 1973) is expressed as the ratio of the luminance of any given sky patch to the zenith luminance. Zenith luminance may then be used to obtain the corresponding absolute quantities.

Hopkinson et al. (1966) and more recently Littlefair (1994) have shown that under dark, overcast skies the Moon and Spencer (1942) proposed CIE overcast sky model shows good agreement with measurements. Under such conditions the zenith luminance may be obtained as

$$L_z = \{9/(7\pi)\}\, IL_G \tag{3.5.22}$$

Perez et al. (1990) have developed an all-sky zenith luminance model which has been tested against extensive data from five locations in North America. The model is represented by

$$L_z / I_D = a_i + b_i \cos z + c_i \exp(-3z) + d_i \Delta \tag{3.5.23}$$

where the coefficients a_i, b_i, c_i and d_i ($i = 1$ to 8) are given in Table 3.5.2.

This model claims to be site and season independent. With an overall bias error around 1% and RMS errors ranging between 0.7 kcd/m² for clear skies and 1.5 kcd/m² for intermediate sky, it represents a significant improvement over the above models. Prog3-4.For may be used to obtain Perez et al. modelled zenith luminance given horizontal diffuse illuminance.

Example 3.5.1

Five minute averaged horizontal global and diffuse irradiance values are provided for Watford, North London (51.71°N, 0.37°W) in Table 3.5.4. Use the Perez et al. model to

Table 3.5.4 Input/output data for Example 3.5.1: North London data, 1 April 1992

Hour	Minute	Input (measured) data					Output (computed) data		
		I_G (W/m²)	I_D (W/m²)	IL_G (klux)	IL_D (klux)	L_Z (Cd/m²)	IL_G (klux)	IL_D (klux)	L_z (Cd/m²)
9	0	483	191	50 578	23 403	3579	51 171	25 086	4061
9	5	502	204	52 334	24 877	3798	52 963	26 523	4399
9	10	516	222	52 788	26 314	4643	54 103	28 468	4869
9	15	547	236	56 389	27 579	4909	57 130	29 971	5250
9	20	521	230	54 087	26 752	3885	54 692	29 396	5132
9	25	386	229	40 863	26 391	4143	41 699	28 011	5381
9	30	392	245	42 205	27 278	6938	42 206	29 763	5865
9	35	286	237	31 695	25 488	9056	31 885	27 121	7385
9	40	186	178	20 320	19 217	6674	21 477	20 652	7757
9	45	144	135	15 939	14 815	4749	17 010	16 013	6242
9	50	352	244	37 223	26 839	6463	39 017	28 798	6452
9	55	455	294	47 040	31 908	8647	49 899	34 238	7830
						MBE	983	1 932	262
						RMSE	171	486	139

Source: data collected by the Building Research Establishment, Garston, UK.

compute horizontal global and diffuse illuminance and the zenith luminance. Evaluate these estimates against the corresponding measured values which are included in Table 3.5.4.

Prog3-4.For may be used to obtain the above estimates for IL_G, IL_D and L_z. These are shown in Table 3.5.4. The MBEs for the horizontal global illuminance indicate an average error of 2.5%. The corresponding error in estimating zenith luminance is 7%.

3.6 Daylight factor

In design studies it has become customary to specify interior daylighting in terms of daylight factor. The daylight factor is the ratio of the internal illuminance to the external illuminance, available simultaneously. It is usually expressed as a percentage. The daylight factor is divided into three components: the direct sky light (sky component), the externally reflected component, and the internally reflected component.

The sky component is the ratio of the illuminance at any given point received from a sky of known luminance distribution to the horizontal illuminance under an unobstructed sky hemisphere. Likewise, the external and internal reflected components are, respectively, the ratios of the illuminance received after reflections from external and internal surfaces to the horizontal illuminance under an unobstructed sky hemisphere. An electronic look-up table for the sky component, based on the CIE standard overcast sky, is provided in Prog3-5.For. This table was originally published by Hopkinson et al. (1966). The BRE protractor and the accompanying literature (BRE, 1986) enable the estimation of the above three components.

It is worth mentioning that Prog3-5.For requires the data file In3-5.Csv which is appended in the CD. The user must load the two files, Prog3-5.For and In3-5.Csv, in a common directory in their PC.

The BRE sky component tables may be used to obtain the daylight level at a reference point in a horizontal plane if the height H and width $2W$ of the window and the distance D of the reference point are known. Figure 3.6.1 shows these details. The electronic look-up table, Prog3-5.For, gives the sky component at the intersections of the ratio H/D and W/D. The geometric construction has to be such that the horizontal and vertical planes drawn through the given point to meet the window wall perpendicularly form two bounding edges of the window. If the window sill is above the reference plane, the height of the sill above the reference plane must also be taken into account. The following example demonstrates the use of Prog3-5.For.

Example 3.6.1

Consider the case of the single-glazed rectangular window shown in Figure 3.6.1. The external obstruction runs along the entire length of the room. Obtain the sky component using Prog3-5.For.

With reference to Figure 3.6.1 we compute the following:

90 SOLAR RADIATION AND DAYLIGHT MODELS

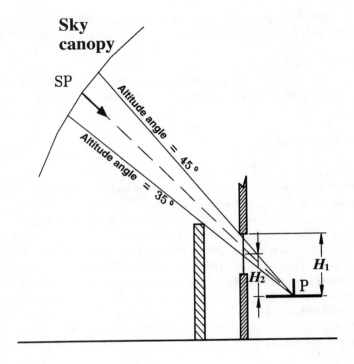

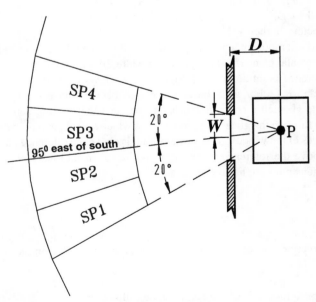

Figure 3.6.1 *Window schematic for Example 3.6.1 (SP sky patch: see section 4.6)*

$W/D = \tan 20° = 0.364 \quad H_1/D = \tan 45° = 1.0 \quad H_2/D = \tan 35° = 0.7$

For the H_1/D, W/D pair Prog3-5.For gives the sky component = 2.02, and for the H_2/D, W/D pair the value of 1.18 is obtained. The actual sky component is then obtained by subtraction of the above two values and then doubling of the result (due to symmetry). Thus

sky component at point P = 2 × 0.84 = 1.68%

It must be borne in mind that the above computation of sky component does not take into account the effects of window orientation, e.g. sun-facing or shaded aspects. It is therefore only an approximate method of obtaining internal illuminance. In Chapter 4 more precise procedures are presented which take into account the window aspect and the real sky luminance distributions. However, owing to its simplicity the above procedure for sky component estimation is widely used.

Hopkinson et al. (1966) have enumerated the advantages of daylight factor as follows. Firstly, it represents the effectiveness of the window as a lighting provider. Secondly, it remains constant even though the outdoor illuminance may fluctuate. Constancy is associated with the concept of adaptation. Appreciation of brightness is governed not only by the actual luminance of the inhabited environment, but also by the brightness of the surroundings which govern the level of visual adaptation. As a result, visual appreciation of the interior of a room does not change radically even though the actual luminance will be higher as a result of the greater amount of daylight penetration coming under brighter skies.

At any given point the daylight factor will result in wide variations in internal illuminance. Acceptance of a given illuminance level as the criterion of an appropriate visual environment poses a problem in relation to the variability of available daylight. One solution is to design in such a way that the recommended level of internal illuminance is attained during a certain agreed proportion of the working period throughout the year. An example of this type of approach is to be found in the work of Hunt (1979) for Bracknell and Kew in the UK. It is possible to undertake illuminance frequency analysis for other locations too using the procedure laid out in the following section.

Figure 3.6.2
Availability of diffuse illuminance for world-wide locations

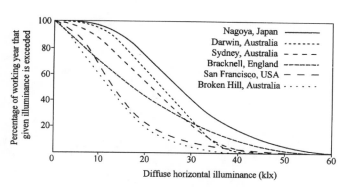

92 SOLAR RADIATION AND DAYLIGHT MODELS

Figure 3.6.2, extracted from Pilkington's (1993) design book, shows the diffuse illuminance frequency distributions for world-wide locations.

3.7 Frequency distribution of irradiation

Prediction of solar energy system performance can be achieved via detailed computer simulations using hourly or sub-hourly weather data or by simpler methods which are based on statistical analysis of long-term measurements. The frequency of occurrence of various levels of daily (and indeed hourly) irradiation is of interest not only from the above viewpoint, but also to enable the determination of the monthly-averaged correlations between diffuse and global quantities. For example, it was shown in Section 2.6 that the latter regression curve may be obtained from the regression for daily diffuse fraction if the frequency distribution of global irradiation during any month was known.

If for any given location the frequency of occurrence of days with various values of K_T is plotted, the resulting distribution would appear like that shown in Figure 3.7.1. This distribution was presented by Hawas and Muneer (1984b), based on long-term measurements for 18 Indian locations. Here it may be seen that the probability of occurrence of K_T between 0.65 and 0.75 is 85%. The shape of this curve depends on the average value of K_T, i.e. lower values of $\bar{K}_T$ would produce curves skewed to the left and vice versa. When the data of Figure 3.7.1 are used to plot cumulative distribution curves, plots such as Figure 3.7.2 are obtained. This figure shows the long-term cumulative frequency f of the daily clearness indices for six Indian locations. It is clear that a set of generalised curves may be obtained for the above locations. As a matter of fact the above technique was originally developed by Liu and Jordan (1963), based on earlier work of Whillier (1956). Later extensions of this work were undertaken by Bendt et al. (1981) for the US, Hawas and Muneer (1984b) for India, and Lloyd (1982) for the United Kingdom. More recently Saunier et al. (1987) have drawn the same conclusions as the latter group of researchers that the above distributions are indeed unique for a given

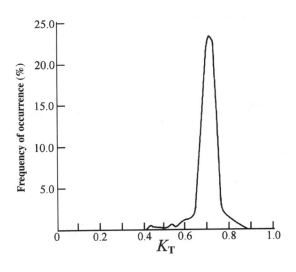

Figure 3.7.1 *Frequency of occurrence of K_T for an Indian location*

Figure 3.7.2
Individual K_T curves for Indian locations

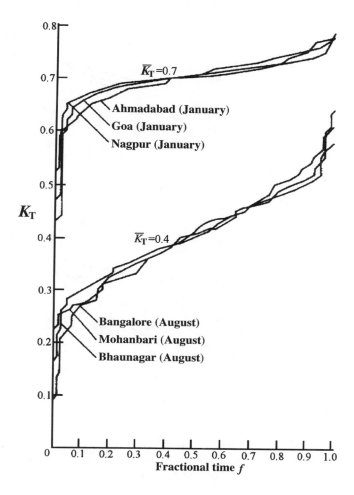

Figure 3.7.3
Comparison of K_T curves for average clearness index 0.3

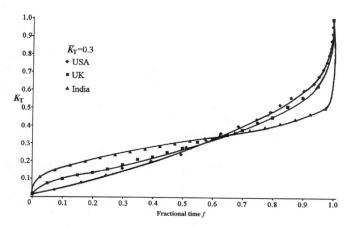

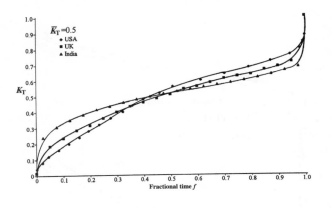

Figure 3.7.4
Comparison of K_T curves for average clearness index 0.5

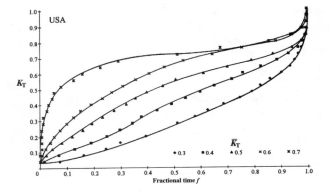

Figure 3.7.5
Generalised K_T curves for USA

geographic area but by no means universal. This point is demonstrated via Figures 3.7.3 and 3.7.4 which show disparate distributions for the US, Indian and UK locations.

Figures 3.7.5–3.7.7 show the K_T distributions, respectively, for the US, India and the UK. These curves, also known as generalised K_T curves, present the insolation character for the respective regions. The curves enable querying the availability of solar energy above a given threshold. For example, for Indian locations, during a month for which $\bar{K}_T = 0.7$, $K_T \leq 0.73$ for 70% fractional time, and $K_T \leq 0.68$ for 20% of the time. In contrast the corresponding figures for the US are, respectively, 56% and 30%. It is therefore evident that the distributions for the Indian locations are flatter. This means that for the Indian subcontinent the daily clearness index K_T varies in a narrower range.

Strictly speaking the above insolation frequency distributions are for daily-based quantities. However, Whillier (1953) has shown that the hourly and daily distribution curves are very similar to each other. Thus the above curves may also be used to obtain the frequency occurrence of hourly global irradiation.

Curve fitting of the above data is quite involved. Attempts have been made by Bendt et al. (1981) to obtain a model using an exponential form. However, as pointed out by Duffie and Beckman (1991), the computations are tedious. For this text, computer look-up tables in FORTRAN have been prepared for ease of use. These are included as Prog3-

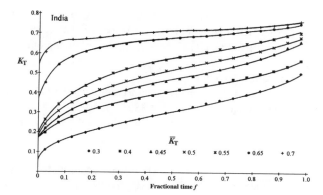

Figure 3.7.6 Generalised K_T curves for India

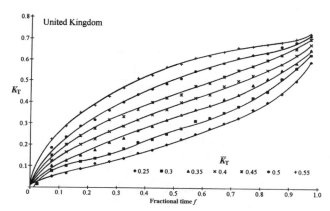

Figure 3.7.7 Generalised K_T curves for the UK

6.For (for the US and North American locations), Prog3-7.For (for Indian and other tropical locations) and Prog3-8.For (for the UK and northern Europe). In these programs the required input is K_T and $\bar{K}_T$, which in turn may be obtained via Prog3-1.For. The following example demonstrates the use of generalised K_T curves.

Example 3.7.1

An office building situated in London is designed to utilise daylight via on–off controls. The working-plane daylight factor is estimated to be 2% and the design illuminance level is set at 500 lx. Using the bright sunshine data for London given in Appendix B, prepare a table of the frequency of occurrence of internal illuminance exceeding the above design value. Perform computations for mid-hour for the normal range of working hours (9–17 hours) for the month of May.

To achieve an internal illuminance of 500 lx, the required external global illuminance would be 25 klx. We assume an approximate and conservative value of 100 lm/W for the global luminous efficacy. Thus, what is required is the frequency of occurrences when horizontal global irradiation exceeds 250 W/m^2.

We note from Appendix B that average daily bright sunshine hours for May are 6.56. Using Prog2-1.For and providing the respective values of 0.13 and, 0.65 for the *a* and *b* coefficients of Eq. (2.1.1) (Table 2.1.2), the following output is obtained for 16 May:

Day length = 15.379 hours
monthly-averaged extraterrestrial irradiation = 10.511 kW h/m^2
averaged global irradiation = 4.28 kW h/m^2
(Page and Lebens (1986) give this value as 4.33 kW h/m^2)
$\bar{K}_T = 0.407$

Now Prog3-3.For and Prog3-8.For may be used to respectively obtain the hourly clearness index K_T and the hourly frequency of occurrence f when the mid-hour irradiance was below the required threshold of 250 W/m^2. These are tabulated as follows:

Hour	K_T	f	f_1	f_2
9.5	0.260	0.245	24.5	75.5
10.5	0.236	0.204	20.4	79.6
11.5	0.225	0.186	18.6	81.4
12.5	0.225	0.186	18.6	81.4
13.5	0.236	0.204	20.4	79.6
14.5	0.260	0.245	24.5	75.5
15.5	0.310	0.330	33.0	67.0
16.5	0.398	0.483	48.3	51.7

Here

f_1 (%) is the frequency of occurrences when internal illuminance is below 500 lx.
f_2 (%) is the frequency of occurrences when internal illuminance exceeds 500 lx.

3.8 Frequency distribution of illuminance

A more direct method for obtaining the frequency distributions of global and diffuse illuminance was presented by Tregenza (1986). In this work the above distributions were derived in relation to solar altitude using empirically developed functions. The estimated cumulative frequency distributions were found to be in good agreement with the measurements carried out at Nottingham (52.9°N) and Garston (51.6°N) in the UK and Uccle (50.8°N) in Belgium.

The procedure to be followed for obtaining the above daylight illuminance distributions (in units of lux) is as follows. Empirical models given by the following equations are used to obtain the mean global and diffuse illuminance:

$$IL_G = 10.5 \, (\text{SOLALT} + 5)^{2.5} \quad -5 < \text{SOLALT} \leq 2.5 \quad (3.8.1a)$$

$$IL_G = 737\,00 \, \sin^{1.22} \text{SOLALT} \quad 2.5 < \text{SOLALT} \leq 60 \quad (3.8.1b)$$

$$IL_D = 10.5 \, (\text{SOLALT} + 5)^{2.5} \quad -5 < \text{SOLALT} \leq 5 \quad (3.8.2a)$$

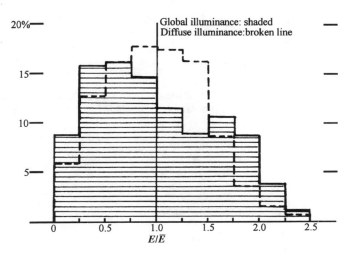

Figure 3.8.1 *Frequency of occurrence of a given horizontal illuminance as a fraction of the mean illuminance*

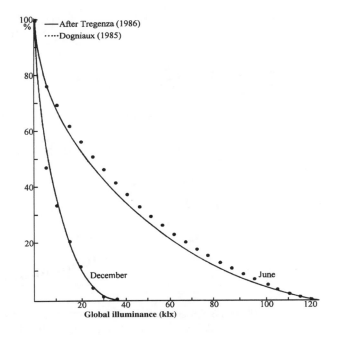

Figure 3.8.2 *Derived cumulative distributions of global illuminance at Uccle for June (18 daylight hours) and December (9.3 hours)*

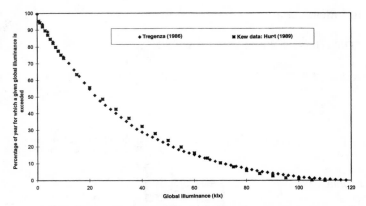

Figure 3.8.3 *Standard working year daylight availability: cumulative global illuminance frequency*

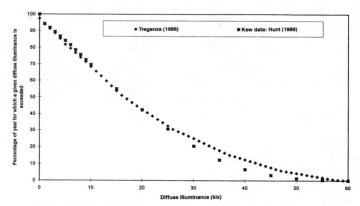

Figure 3.8.4 *Standard working year daylight availability: cumulative diffuse illuminance frequency*

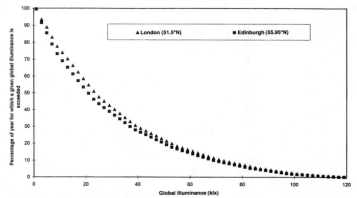

Figure 3.8.5 *Standard working year daylight availability: cumulative global illuminance frequency, London and Edinburgh*

$IL_D = 48\,800\,\sin^{1.105} \text{SOLALT} \quad 5 < \text{SOLALT} \leq 60$ (3.8.2b)

The solar altitude is then calculated (using say Prog1-6.For) on intervals during the day and regular dates throughout the year. At each time the mean illuminance is then computed from the above models. The relative frequencies of illuminance of 0.125, 0.375, 0.625, 0.875, . . ., 2.375 times the mean are obtained from the values illustrated in Figure 3.8.1. At each calculated instance ten values of frequencies are obtained which may be treated to be statistically similar to time-based measurements.

The above procedure has the advantage that it extends its application to daylight as well as the civil twilight band. Figure 3.8.2 compares the estimated illuminance frequency for Uccle using the method under discussion with those obtained by Dogniaux (1985).

Hunt (1979) has analysed several years of records from Kew, Bracknell and five other UK locations to produce graphs and tables of the frequency distribution of horizontal global and diffuse illuminance. This type of information is a useful aid for daylight designers for exploring the potential for energy savings via modern photoelectric controls (see Figure 3.5.1). Figures 3.8.3 and 3.8.4 provide a comparison of Tregenza's (1986) method and the data presented by Hunt (1979) for Kew. The agreement is highly satisfactory for the global illuminance. Reasonably good agreement may also be noted, bar the intermediate-level illuminance, for the diffuse component. It is well known that the diffuse illuminance from a part-overcast sky exhibits the most pronounced scatter.

Using the procedure outlined in the present and the preceding sections it is therefore possible to obtain the frequency of occurrence of a given daylight level for other locations. One such plot, obtained via Prog3-9.For, is shown in Figure 3.8.5. Prog3-9.For performs stepwise computation of:

(a) solar altitude at each 15 minute increment for each day of the year
(b) the relative occurrence of illuminance equal to 0.125, 0.375, . . ., 2.375 times the mean
(c) the cumulative frequency distribution of horizontal global and diffuse illuminance.

The only input required to execute the above routine is the latitude and longitude of the location for which the above illuminance distributions are required. For any geographic region, subject to the availability of fractional mean frequencies such as those shown in Figure 3.8.1 and Table 3.8.1, the following daylight design tables and charts may be produced:

(a) global and diffuse illuminance frequency distributions
(b) occurrence of a given illuminance level on a monthly or mean hourly basis
(c) effect of latitude
(d) effect of summer time shift, office and school operating hours and lunch breaks on daylight use.

The sociological and energetic implications of a shift in the clock time are significant. In the UK, for instance, an experiment of maintaining British Summer Time (BST)

Table 3.8.1 Frequency of occurrence of a given horizontal illuminance*
(after Tregenza, 1986)

Fraction of mean illuminance†	Frequency of occurrence	
	Global	Diffuse
0.125	8.87	5.96
0.375	15.78	12.86
0.625	16.15	16.19
0.875	14.67	17.65
1.125	11.45	17.39
1.375	8.87	16.27
1.625	10.62	8.63
1.875	8.68	3.55
2.125	3.79	0.80
2.375	1.12	0.70

* Any irregularities in the frequency of measured occurrences may be due to the conditions in the relevant period. Longer-term data may be represented by a smoother curve.
† Mid-point of frequency band.

through the winter was conducted from 1968 to 1971. A study of the associated costs and benefits of that experiment were published by the Policy Studies Institute (1988). A rigorous and updated review of this subject was carried out by Hillman (1993). The latter work examines the effects of Double British Summer Time (DBST) and of putting the clock forward from the end of summer to the beginning of spring. It has been shown that such a change would have a direct bearing on road accidents, security, health and leisure, and agricultural activities as well as the fuel consumption due to lighting of domestic and office buildings. The tools presented in this section, e.g. Prog3-9.For, enable such impacts to be quantified.

3.9 Exercises

3.9.1 Ten-year monthly-averaged data for Bracknell, UK are given in Tables 3.3.6 and 3.3.7. Using Liu–Jordan's regression models (3.1.1) and (3.2.1), obtain hourly values of global and diffuse irradiation from monthly-averaged daily data (Table 3.3.7). Compare your estimates against the MRM predictions which are provided in Table 3.3.6.

3.9.2 Using the sunshine data for Stornoway and Bracknell, UK given in Table 3.3.7, obtain monthly-averaged daily estimates of global irradiation using Eq. (2.1.1). You may then proceed to obtain monthly-averaged daily diffuse irradiation using Eq. (2.2.2). Compare these with the estimates of MRM given in Table 3.3.7.

3.9.3 Hourly synoptic (weather) and irradiance data for Easthampstead, UK (51.383°N) are given in Table 3.9.1. Prepare the input file In3-2.Csv containing the data

Table 3.9.1 Synoptic and radiation data for Easthampstead, UK (51.38°N), June 1991

Day	Hour	Dry-bulb temp. (°C)	Wet-bulb temp. (°C)	Atmosphere pressure (mbar)	SF	I_G (W/m²)	I_D (W/m²)	$I_{G,TLT}$ (W/m²)	$I_{G,N}$ (W/m²)	$I_{G,E}$ (W/m²)	$I_{G,S}$ (W/m²)	$I_{G,W}$ (W/m²)
11	0	0.0	0.0	1018.0	0.0	0	0	0	0	0	104	93
11	1	0.0	0.0	1018.5	0.0	0	0	0	0	0	95	87
11	2	0.0	0.0	1018.8	0.0	0	0	0	0	0	82	77
11	3	0.1	0.1	1019.0	0.0	1	1	0	2	0	82	79
11	4	5.0	3.1	1018.9	0.2	23	106	14	181	15	84	80
11	5	17.7	6.7	1018.9	1.0	57	217	34	534	32	89	85
11	6	31.7	7.8	1019.1	1.0	152	142	47	685	38	100	93
11	7	41.2	22.2	1019.1	0.5	301	112	115	516	79	120	105
11	8	32.2	27.2	1019.5	0.2	257	107	131	199	97	140	114
11	9	23.8	22.5	1019.3	0.0	186	91	96	118	83	150	110
11	10	23.4	23.2	1019.6	0.0	173	96	88	96	89	150	110
11	11	33.9	32.7	1019.5	0.0	260	121	133	122	130	139	116
11	12	40.4	39.0	1018.9	0.0	326	139	175	157	147	130	104
11	13	24.9	24.5	1017.9	0.0	187	100	96	102	95	148	113
11	14	25.0	24.7	1017.4	0.0	188	98	96	100	98	152	114
11	15	19.8	19.5	1017.1	0.0	148	78	75	84	72	148	114
11	16	13.2	13.1	1016.7	0.0	93	56	46	55	50	145	116
11	17	8.6	8.5	1016.0	0.0	62	35	31	33	34	146	118
11	18	2.8	2.7	1015.7	0.0	18	11	9	11	10	142	115
11	19	1.1	1.0	1014.8	0.0	7	4	3	3	3	141	117
11	20	0.1	0.1	1014.4	0.0	0	0	0	0	0	137	120
11	21	0.0	0.0	1013.9	0.0	0	0	0	0	0	135	120
11	22	0.0	0.0	1013.2	0.0	0	0	0	0	0	134	122
11	23	0.0	0.0	1012.4	0.0	0	0	0	0	0	132	125
12	0	0.0	0.0	1011.4	0.0	0	0	0	0	0	130	124
12	1	0.0	0.0	1010.8	0.0	0	0	0	0	0	130	124
12	2	0.0	0.0	1010.2	0.0	0	0	0	0	0	130	124
12	3	0.0	0.0	1009.4	0.0	0	0	0	0	0	128	123
12	4	1.3	1.3	1008.6	0.0	9	5	4	5	5	128	123
12	5	2.9	2.9	1008.4	0.0	21	12	10	11	11	129	124
12	6	6.4	6.3	1007.5	0.0	46	27	23	27	25	132	126
12	7	19.4	18.0	1007.5	0.2	150	76	77	126	61	135	129
12	8	43.6	24.1	1007.3	0.8	379	95	181	397	70	145	134
12	9	61.5	25.5	1007.3	1.0	570	102	290	432	89	159	135
12	10	55.7	35.3	1006.9	1.0	503	128	270	264	125	170	131
12	11	77.3	29.1	1006.8	1.0	743	112	399	199	126	176	130
12	12	51.2	31.2	1006.2	1.0	478	114	262	132	168	184	134
12	13	21.3	19.4	1005.5	0.1	179	79	98	86	83	195	136
12	14	18.5	17.3	1005.2	0.2	129	77	61	62	98	185	136
12	15	33.2	20.2	1005.4	0.4	281	76	136	80	237	165	130
12	16	15.0	14.1	1004.8	0.2	118	54	62	59	74	175	135
12	17	7.9	7.3	1004.7	0.0	56	38	29	28	59	165	130
12	18	13.9	9.7	1005.0	0.5	68	123	38	38	258	155	122
12	19	2.6	2.5	1005.2	0.0	20	15	11	11	17	144	120
12	20	0.1	0.1	1005.2	0.0	1	0	0	0	0	141	113
12	21	0.0	0.0	1005.4	0.0	0	0	0	0	0	139	113
12	22	0.0	0.0	1005.6	0.0	0	0	0	0	0	129	108
12	23	0.0	0.0	1005.3	0.0	0	0	0	0	0	124	111

$I_{G,TLT}$: irradiance on a south-facing surface at the local latitude angle.
$I_{G,N}, I_{G,E}, I_{G,S}, I_{G,W}$: irradiance on a north-, east-, south- and west-facing vertical surface.

fields for year, month, day, hour, dry-bulb temperature, wet-bulb temperature, atmospheric pressure and sunshine fraction (columns 3, 4, 5 and 6). Hence execute Prog3-2.For (or Prog3-2.Exe) to obtain hourly horizontal global and diffuse irradiation (columns 7 and 8). Compare your estimates against the measured data which are included in Table 3.9.1. Prepare the MBE and RMSE statistics for your estimates. (Hint: a sample electronic file In3-2.Csv is available on the CD.)

Save your results for the computation of slope irradiation (see Exercise 4.7.3) from your estimated values of horizontal irradiation. (Hint: you may use a spreadsheet medium, such as Excel, to prepare file In3-2.Csv and to import Out3-2.Dat. The MBE and RMSE statistics may then be obtained in Excel.)

3.9.4 Use Prog3-3.For, based on Eqs (3.4.1) and (3.4.2), to compute hourly horizontal diffuse irradiation using the global irradiation values given in Table 3.9.1. Comment on the accuracy of the computed estimates against those obtained in Exercise 3.9.3.

3.9.5 Make appropriate modifications to Prog3-3.For so that Eq. (3.4.3) is substituted for Eq. (3.4.2). Execute this program after estimating the hourly relative humidity data from the corresponding dry-bulb and wet-bulb temperatures using Prog7-1.Exe. Compare the accuracy of the diffuse irradiation estimates obtained from the three models, i.e. Eqs (3.4.1), (3.4.2) and (3.4.3).

3.9.6 Use Prog3-4.For, based on the Perez et al. (1990) luminous efficacy model, to compare its performance against the 5 minute averaged (measured) data provided in File3-1.Csv. Recall that this file contains Edinburgh data for April 1993.

3.9.7 Compare the accuracy of the above estimates against the DuMortier–Perraudeau Page model presented in Section 3.5.6.

3.9.8 Refer to Figure 3.6.1 and Example 3.6.1. Assume that the distance of the reference point at which the sky component was computed is 2.75 m from the window. Use Prog3-5.For to obtain the sky components successively for points which are positioned at linear distances of 0.5 m, 1 m and 1.5 m away from the original position shown in Figure 3.6.1.

3.9.9 Using Prog3-8.For and the monthly-averaged data provided in Table 3.1.1, obtain the frequency distribution of global irradiance and illuminance for 10 and 11 hours. (Hint: you may use an average global luminous efficacy of 110 lm/W.)

3.9.10 Compare the above frequency distribution of global illuminance against the distribution computed via the Tregenza model given in Section 3.8. Comment on any differences you may notice between the two frequency distributions.

References

Ångström, A. (1929) On the atmospheric transmission of sun radiation and on dust in the air. *Geografis. Annal.* 2, 156–66.
Ångström, A. (1930) On the atmospheric transmission of sun radiation. *Geografis. Annal.* 2 and 3, 130–59.
Aydinli, S. and Krochmann, J. (1983) *Data on daylight and solar radiation: Guide on daylight*. Draft for CIE TC 4.2, Commission Internationale de l'Éclairage, Paris.
Barbaro, S., Coppolino, S., Leone, C. and Sinagra, E. (1979) An atmospheric model for computing direct and diffuse solar radiation. *Solar Energy* 22, 225.
Bendt, P., Collares-Pereira, M. and Rabl, A. (1981) The frequency distribution of daily radiation values. *Solar Energy* 27, 1.
Bird, R.E. and Hulstrom, R.L. (1979) *Application of Monte Carlo Technique to Insolation Characterisation and Prediction*. US SERI Tech. Report TR-642-761, 38.
Bird, R.E. and Hulstrom, R.L. (1981) *A Simplified Clear-Sky Model for the Direct and Diffuse Insolation on Horizontal Surfaces*. US SERI Tech. Report TR-642-761, 38.
Blackwell, M.J. (1954) *Five Years' Continuous Recordings of Total and Diffuse Solar Radiation at Kew Observatory*. Met. Res. Publication 895, Met. Office, London.
Boes, E.C., Hall, I.J., Prarie, R.R., Stromberg, R.P. and Anderson, H.E. (1976) *Report SAND 76-0411*. Sandia Nat. Lab., Albuquerque, NM.
BRE (1986) *Estimating Daylight in Buildings: Parts I and II*. BRE Digests 309, 310. Building Research Establishment, Watford.
Bruno, R. (1978) A correction procedure for separating direct and diffuse insolation on a horizontal surface. *Solar Energy* 20, 97.
Bugler, J.W. (1977) The determination of hourly insolation on an inclined plane using a diffuse irradiance model based on hourly measured global horizontal. *Solar Energy* 19, 477.
Building Services and Environmental Engineer (1996) News International. March, p. 5.
CEC (1996) *European Solar Radiation Atlas*, Eds. W. Palz and J. Grief. Springer-Verlag, Berlin. Commission of the European Communities.
Chandrasekhar, S. (1950) *Radiative Transfer*. Oxford University Press, London.
Chandrasekhar, S. and Elbert, D.D. (1954) The illumination and polarization of the sunlit sky on Rayleigh scattering. *Trans. Amer. Phil. Soc.* 44, 643.
Chauvel, P. (1993) *Dynamic Characteristics of Daylight Data and Daylighting Design in Buildings*. Final Report JOUE CT 90-0064, CEC, Brussels.
Chroscicki, W. (1971) Calculation methods of determining the value of daylight intensity on the ground of photometric and actinometric measurements. *Proc. CIE Barcelona Conf.* 71.24.
CIBSE (1996) *Building Services Supplement,* May 1996. Chartered Institution of Building Services Engineers, London.
CIBSE (1997) *CIBSE Guide J: Weather and Solar Data*. Chartered Institution of Building Services Engineers, London.
CIE (1973) *Standardization of Luminance Distribution of Clear Skies*. CIE no. 22 (TC-4.2). Commission Internationale de l'Éclairage, Paris.
Collares-Pereira, M. and Rabl, A. (1979) The average distribution of solar radiation –

correlations between diffuse and hemispherical and between daily and hourly insolation values. *Solar Energy* 22, 155.

Coulson, K.L. (1959) Characteristics of radiation emerging from the top of a Rayleigh atmosphere. *Planet Space Sci.* 1, 265.

Dave, J.V. (1964) Importance of higher order scattering in a molecular atmosphere. *J. Opt. Soc. Amer.* 54, 307.

Dave, J.V. (1979) Extensive data sets of the diffuse radiation in realistic atmospheric models with aerosols and common absorbing gases. *Solar Energy* 21, 361–9.

Davies, J.A., Schertzer, W. and Nunez, M. (1975) Estimating global solar radiation. *Boundary-Layer Meteorol.* 9, 33–52.

Delaunay, J.J. (1995) *Development and Performance Assessment of Luminous Efficacy Models*. Internal Report, Fraunhofer Institute for Solar Energy Systems, Freiburg, Germany.

Dogniaux, R. (1960) Données météorologiques concernant l'ensoleillement et l'éclairage naturel. *Cah. Cent. Sci. Batim.* 44, 24.

Dogniaux, R. (1979) *Variations qualitatives et quantitatives des composante du rayonment solaire sur une surface horizontale par ciel serein enfonction du trouble atmosphérique*. Publication IRM, Serie b(62), Institut Royal Météorologique de Belgique, Brussels.

Dogniaux, R. (1985) *Disponibilité de la lumière du jour*. Institut Royal Météorologique de Belgique, Brussels.

Drummond, A.J. (1958) Notes on the measurement of natural illumination II. Daylight and skylight at Pretoria: the luminous efficacy of daylight. *Arch. Met. Wien*, B9, 149.

Drummond, A.J. and Ångström, A.K. (1971) Derivation of the photometric flux of daylight from filtered measurements of global (sun and sky) radiant energy. *Appl. Opt.* 10, 2024.

Duffie, J.A. and Beckman, W.A. (1991) *Solar Engineering of Thermal Processes*. Wiley, New York.

DuMortier, D. (1994a) *Modelling Global and Diffuse Horizontal Irradiances under Cloudless Skies with Different Turbidities*. Report for JOULE 2 Project, CEC, Brussels.

DuMortier, D. (1994b) *Discussion on the Prediction of Irradiances for Clear Sky Conditions*. Report for JOULE 2 Project, CEC, Brussels.

Erbs, D.G., Klein, S.A. and Duffie, J.A. (1982) Estimation of the diffuse fraction of hourly, daily and monthly average global radiation. *Solar Energy* 28(6), 293–302.

Goody, R.M. (1964) *Atmospheric Radiation I: Theoretical Basis*. Oxford University Press, London.

Grindley, P.C., Batty, W.J. and Probert, S.D. (1995) Mathematical model for predicting the magnitudes of total, diffuse and direct-beam insolation. *Applied Energy* 52, 89.

Gueymard, C. (1993) Critical analysis and performance assessment of clear sky solar irradiance models using theoretical and measured data. *Solar Energy* 51, 121.

Gueymard C. (1995) SMARTS2, *a Simple Model of the Atmospheric Radiative Transfer of Sunshine: Algorithms and Performance Assessment*. Report FSEC-PF-270-95, Florida Solar Energy Center, Cocoa, FL.

Hawas, M. and Muneer, T. (1984a) Study of diffuse and global radiation characteristics in India. *En. Conv. & Mgmt* 24, 143.

Hawas, M. and Muneer, T. (1984b) Generalized monthly K_T-curves for India. *En. Conv. & Mgmt* 24, 185.

Hay, J.E. (1976) A revised method for determining the direct and diffuse components of the total shortwave radiation. *Atmosphere* 14, 278.

Hay, J.E. (1979) Calculation of monthly mean solar radiation for horizontal and inclined surfaces. *Solar Energy* 23, 301.

Hillman, M. (1993) *Time for a Change – a New Review of the Evidence*. Policy Studies Institute, London.

Hopkinson, R.G., Petherbridge, P. and Longmore, J. (1966) *Daylighting*. Heinemann, London.

Hunt, D.R.G. (1979) Availability of Daylight. Building Research Establishment, Garston.

Iqbal, M. (1979) Prediction of hourly diffuse solar radiation from measured hourly global radiation on a horizontal surface. *Solar Energy* 24, 491.

Iqbal, M. (1983) *An Introduction to Solar Radiation*. Academic Press, New York.

Jeter, M. and Balaras, C.A. (1990) Development of improved solar radiation models for predicting beam transmittance. Solar Energy 44, 149.

Kambezidis, H., Psiloglou, B. and Synodinou, B. (1997) Comparison between measurements and models for daily solar irradiation on tilted surfaces in Athens, Greece. *J. of Renewable Energy* (In Press).

Karayel, M., Navvab, M., Neeman, E. and Selkowitz, S. (1983) Zenith luminance and sky luminance distribution for daylighting applications. *Energy and Buildings* 6, 3.

Kasten, F. (1993) Discussion on the relative air mass. *Lighting Res. & Tech.* 25, 129.

Kittler, R. (1970) *Some Considerations Concerning the Zenith Luminance of the Cloudless Sky*. Circular no. 11, CIE E-3.2.

Koga, Y., Nakamura, H. and Rahim, M.R. (1993a) Study on luminous efficacy – The relation to cloud cover. *Proc. Lux Europa 1993*, Edinburgh, 4–7 April 1993, vol. II, pp. 799–803.

Koga, Y., Nakamura, H. and Rahim, M.R. (1993b) Daylight and solar radiation data in Ujung Pandang, Indonesia. *Proc. Second Lux Pacifica Lighting Conf.*, Bangkok, 10–13 November 1993, pp. C85–C90.

Kondratyev, K.Y. (1969) *Radiation in the Atmosphere*. Academic Press, New York.

Krochmann, J. (1970) Uber die horizontal beleuchtungs-starke und die zenitleuchtdichte des klaren himmels. *Lichttechnik* B22, 551.

Lacis, A.A. and Hansen, J.E. (1974) A parameterisation for the absorption of solar radiation in the earth's atmosphere. *J. Atmos. Sci.* 31, 118–132.

Liebelt, C. (1975) Leuchtdichte- und Strahldichteverteilung durch Tageslicht. *Gesundsheitsingenieur* 96, 127.

Littlefair, P.J. (1985) The luminous efficacy of daylight: a review. *Lighting Res. & Tech.* 17, 162.

Littlefair, P.J. (1988) Measurement of the luminous efficacy of daylight. *Lighting Res. & Tech.* 20, 177.

Littlefair, P.J. (1994) A comparison of sky luminance models with measured data from Garston, United Kingdom. *Solar Energy* 53, 315.

Littlefair, P.J. (1996) Internal Report. Building Research Establishment, Garston, UK.

Liu, B.Y.H. and Jordan, R.C. (1960) The inter-relationship and characteristic distribution of direct, diffuse and total solar radiation. *Solar Energy* 4, 1.

Liu, B.Y.H. and Jordan, R.C. (1963) The long-term average performance of flat plate solar energy collectors. *Solar Energy* 7, 53.

Lloyd, P.B. (1982) *A Study of Some Empirical Relations Described by Liu and Jordan.* Report no. 333, Solar Energy Unit, University College, Cardiff.

Lunde, P.J. (1980) *Solar Thermal Engineering*, Wiley, New York.

Mani, A. and Rangarajan, S. (1983) Techniques for the precise estimation of hourly values of global, diffuse and direct solar radiation. *Solar Energy* 31, 577.

Maxwell, E.L. (1987) *A Quasi-Physical Model for Converting Hourly Global Horizontal to Direct Normal Insolation.* Report SERI/TR-215-3087. Solar Energy Research Institute, Golden, Co.

McDermott, L.H. and Gordon-Smith, G.W. (1951) Daylight illumination recorded at Teddington. *Proc. Build. Res. Congr.,* Division 3, Part III, 156.

Meteorological Office (1980) *Solar Radiation Data for the United Kingdom.* MO 912, Meteorological Office, Bracknell.

Moon, P. (1940) Proposed standard solar radiation curves for engineering use. *J. Franklin Inst.* 230, 583.

Moon, P. and Spencer, D.E. (1942) Illumination from a non-uniform sky. *Trans. Illum. Eng. Soc.,* 37, 707.

Muneer, T. (1987) *Solar Radiation Modelling for the United Kingdom.* PhD thesis, Council for National Academic Awards, London.

Muneer, T. (1995) Solar irradiance and illuminance models for Japan II. Luminous efficacies. *Lighting Res. & Tech.* 27, 223.

Muneer, T. and Angus, R.C. (1993) Daylight illuminance models for the United Kingdom. *Lighting Res. & Tech.* 25, 113.

Muneer, T. and Angus, R.C. (1995) Luminous efficacy: evaluation of models for the United Kingdom. *Lighting Res. & Tech.* 27, 71.

Muneer, T. and Gul, M. (1997) Detailed evaluation of a meteorological radiation model using long-term data from UK locations. *En. Conv. & Mgmt* (In Press).

Muneer, T., Gul, M., Kambezedis, H. and Alwinkle, S. (1996) An all-sky solar radiation model based on meteorological data. *Proc. CIBSE / ASHRAE Joint Annual Conf.,* Harrogate.

Muneer, T, Hawas, M.M. and Sahili, K. (1984) Correlation between hourly diffuse and global radiation for New Delhi. *En. Conv. & Mgmt* 24, 265.

Muneer, T. and Saluja, G.S. (1986) Correlation between hourly diffuse and global solar radiation for the UK. *BSER&T* 7, 37.

Nagata, T. (1970) Luminance distribution of clear skies, Part 2. Theoretical considerations. *Trans. Arch. Inst. Japan* 186, 41.

Nakamura, H. and Oki, M. (1986) The mean sky composition, its dependence on the absolute luminance of the sky elements and its application to the daylighting prediction. *Proc. Int. Daylighting Conf.,* Long Beach, California, p. 61.

Navvab, M., Karayel, M., Neeman, E. and Selkowitz, S. (1988) Luminous efficacy of daylight. *Proc. CIBSE Nat. Light. Conf.,p. p.*409.

Orgill, J.F. and Hollands, K.G.T. (1977) Correlation equation for hourly diffuse radiation on a horizontal surface. *Solar Energy* 19, 357.

Page, J.K. and Lebens, R. (1986) *Climate in the United Kingdom.* HMSO, London.

Perez, R., Ineichen, P. and Seals, R. (1990) Modelling daylight availability and irradiance components from direct and global irradiance. *Solar Energy* 44, 271–89.

Perez, R., Ineichen, P., Maxwell, E., Seals, R. and Zelenka, A. (1991) Dynamic global-to-direct irradiance conversion models. *Proc. ISES World Congress,* Denver, Co.

Pilkington (1993) *Glass in Buildings,* eds D. Button and B. Pye. Pilkington PLC, Prescott Road, St Helens, United Kingdom.

Pisimanis, D., Notaridou, V. and Lalas, D.P. (1987) Estimating direct, diffuse and global radiation on an arbitrarily inclined plane in Greece. *Solar Energy* 39 (3), 159.

Pleijel, G. (1954) The computation of natural radiation in architecture and town planning. *Meddelande Bull., Statens Namnd for Byggnadsforskning, Stockholm* 25, 30.

Policy Studies Institute (1988) Making the Most of Daylight Hours. Policy Studies Institute, London.

Rahim, M.R., Nakamura, H., Koga, Y. and Matsuzawa, T. (1993) The modified equation for the zenith luminance of the clear sky. Proc. Second Lux Pacifica Lighting Conf., Bangkok, 10–13 November 1993.

Red Book (1992) *Ozone Data for the World.* Atmospheric Environment Service, Downsview, Ontario, Canada, in cooperation with the World Meteorological Organisation, 33, no. 6, November–December 1992.

Reindl, D.T., Beckman, W.A. and Duffie, J.A. (1990) Diffuse fraction correlations. *Solar Energy* 45, 1–7.

Reitan, C.H. (1963) Surface dew-point and water vapor aloft. *J. Appl. Meteor.* 2, 776.

Saluja, G.S. and Robertson, P. (1983) Design of passive solar heating in the northern latitude locations. *Proc. Solar World Congress*, Perth, Australia, p. 40.

Saunier, G.Y., Reddy, T.A. and Kumar, S. (1987) A monthly probability distribution function of daily global irradiation values appropriate for both tropical and temperate locations. *Solar Energy* 38, 169.

Sekera, Z. (1956) Recent developments in the study of the polarization of skylight. *Adv. Geophys.* 3, 43.

Shettle, E.P. and Fenn, R.W. (1975) Models of the atmospheric aerosol and their optical properties. *Proc. AGARD Conf. no. 183 Optical Propagation in the Atmosphere,* pp. 2.1–2.16.

Spencer, J.W. (1982) Correlation equation for hourly diffuse radiation on a horizontal surface. *Solar Energy* 29, 19.

Tregenza, P.R. (1980) A simple mathematical model of illumination from a cloudy sky. *Lighting Res. & Tech.* 12, 121.

Tregenza, P.R. (1986) Measured and calculated frequency distributions of daylight illuminance. *Lighting Res. & Tech.* 18, 71.

US Standard Atmosphere (1976) US Government Printing Office, Washington, DC.

Van Heuklon, T.K. (1979) Estimating atmospheric ozone for solar radiation models. *Solar Energy* 22, 63–8.

Whillier, A. (1953) *Solar Energy Collection and its Utilization for House Heating.* PhD thesis, MIT, Cambridge, MA.

Whillier, A. (1956) The determination of hourly values of total radiation from daily summations. *Arch. Met. Geoph. Biokl.* B7, 197.

Wright, J., Perez, R. and Michalsky, J.J. (1989) Luminous efficacy of direct irradiance: variations with insolation and moisture conditions. *Solar Energy* 42, 387.

4 HOURLY SLOPE IRRADIATION AND ILLUMINANCE

In the previous chapters, models for incident horizontal diffuse and global irradiation and illuminance were discussed. Having obtained these quantities, the next step is to obtain the incident slope beam and diffuse energy. Diffuse irradiance or illuminance on any sloping surface consists of sky-diffuse and ground-reflected components. While the latter will be discussed at length in Chapter 6, the sky-diffuse component will be considered here.

The task of computing beam (direct) energy is a matter related to solar geometry. Computation of the diffuse component on a surface of given orientation and tilt is, however, not as simple. Chapter 1 provides the equations and FORTRAN programs related to solar geometry. Most programs in this chapter, which enable computation of the slope beam, sky-diffuse and isotropic ground-reflected components, have been developed on the Prog1-6.For platform which is a convenient and precise routine for solar geometry.

Historically, the development of sky-diffuse models initiated with the work of Moon and Spencer (1942) who used measured data from Kiel in Germany, and Chicago and Washington DC in the USA to demonstrate the anisotropic nature of the luminance distribution of overcast skies. Figure 4.0.1 presents their findings.

The irradiance model development has, however, been rather slow, e.g. even in the

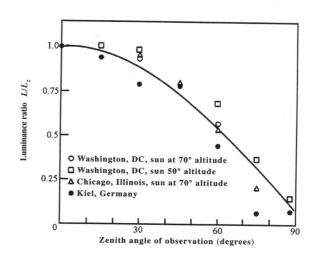

Figure 4.0.1 *Luminance distribution for overcast skies (Moon and Spencer, 1942)*

early seventies Liu and Jordan's (1960) isotropic model was employed in very many studies. In the present chapter the irradiance model development is categorised in three sections, referred to as the first, second and third generation models. The development of slope illuminance models will be examined in Section 4.4. The radiance and luminance distributions of the sky have evolved from a 'discrete' to a 'continuous' mode and these will be presented in Section 4.5.

Dubayah and Rich (1995) have presented an overview of the effects of topography and plant canopies on solar radiation. One of the important aspects of that work is the presentation of the effects of shade and shadowing on the land mass surrounding hills. The corresponding impact on the biophysical processes, soil heating and water balances was also presented. It will be shown herein that new developments in slope irradiance modelling enable such intricate details to be incorporated in large scale applications such as geometeorology as well as building design.

4.1 Slope beam irradiance and illuminance

As shown in Section 2.7, the daily slope beam irradiation is obtained from the horizontal beam irradiation by multiplying it with the beam conversion factor R_B. Similarly, the hourly or instantaneous slope beam irradiance is obtained via

$$I_{B, \text{TLT}} = I_B \, r_B \tag{4.1.1a}$$

$$r_B = \max[0, \cos \text{INC}/\sin \text{SOLALT}] \tag{4.1.1b}$$

I_B is the difference between the hourly, or instantaneous, horizontal global I_G and diffuse irradiance I_D.

4.2 Sky clarity indices

Many anisotropic sky-diffuse models use one form or another of sky clarity index to describe the prevailing condition. Under non-overcast conditions the constituent components of sky-diffuse irradiance are a circumsolar (sun's aureole) part and background diffuse irradiance. The sky clarity indices are used to 'mix' the above mentioned components. Some of these indices will be presented herein.

The most common of all the sky clarity indices is the 'clearness function' F, given by

$$F = (I_G - I_D)/I_E \tag{4.2.1}$$

I_E is the horizontal extraterrestrial irradiance (W/m²). Eq. (3.3.7) and Prog4-1.For enable computation of I_E.

The index proposed by Klucher (1979) is of the form

$$F' = 1 - (I_D/I_G)^2 \tag{4.2.2}$$

It may also be recalled that in Chapter 3 the hourly clearness index, $K_t = I_G/I_E$ was introduced.

4.3 Sky-diffuse irradiance models

With the exception of clear skies, very often the sky-diffuse irradiance is the dominant component. Precise estimation of it is therefore important for all solar energy related work. The three sub-sections presented below, though not exhaustive, outline the historical development of the relevant models. A FORTRAN program which enables computation of instantaneous slope beam, sky-diffuse and ground-reflected irradiance is provided in Prog4-1.For. The program estimates sky-diffuse irradiance for seven selected models, the selection being made to reflect their general currency and rigorous validations undertaken in the past, e.g. Kambezidis et al. (1994).

4.3.1 First generation models

As the name suggests, these are the earliest and simplest of all models. Although easy to use, owing to their imprecise nature they are increasingly being replaced by more sophisticated second and third generation models.

4.3.1.1 Isotropic model
The simplest of all slope irradiance models is the one which assumes an isotropic sky. The following equation may easily be derived using the radiation configuration analysis:

$$I_{D, \text{TLT}} = I_D \cos^2 (\text{TLT}/2) \tag{4.3.1.1}$$

However, diffuse irradiation is not isotropic in nature and is an angular function of the solar altitude and azimuth. Kondratyev and Manolova (1960) have made an excellent study of the nature of diffuse and reflected radiation. They measured the radiation intensity in 37 directions, for the tilt angles of 15°, 40° and 65° in every 30° of azimuth and the zenith. The anisotropic nature of sky radiance was thus demonstrated.

The following are the broad conclusions drawn from the study of Kondratyev and Manolova:

(a) Measurements of diffuse and reflected irradiation show that the distribution is a function of the azimuth and solar altitude.
(b) In the case of overcast skies (dense cloudiness) the isotropic condition proves to be satisfactory.
(c) However, when the clouds are not uniform and partial transparency exists, the isotropic condition is unrealistic.

4.3.1.2 Circumsolar model
Another simple model which belongs to the first generation set is the circumsolar model, which assumes that the sky-diffuse radiation, as well as the beam radiation, emanates from

the direction of the solar disk (Iqbal, 1983). Mathematically, this may be expressed as

$$I_{D, TLT} = I_D r_B \tag{4.3.1.2}$$

The above model may only be adopted under exceptionally clear sky conditions. Even under those conditions it will only approximately represent the radiance distribution and, as such, has never been a serious contender.

4.3.2 Second generation models

These models differentiate between the radiance distribution of clear and overcast skies. However, they do not completely divorce their generic development from the isotropic sky and as such revert to the latter under overcast conditions. In terms of accuracy they offer a fair improvement over the first generation models.

4.3.2.1 Temps and Coulson's model

Temps and Coulson (1977) have suggested an anisotropic modification to the clear sky diffuse radiance model. They used clear sky measurements to demonstrate the anisotropic nature of diffuse irradiation.

The findings of Temps and Coulson for clear sky radiance distribution may be summarised as follows:

(a) The main deficiency in the isotropic model seems to be due to the fact that an increased intensity exists near the horizons and in the circumsolar region of the sky.
(b) Observations showed that skylight intensity is about 40% greater at horizons than at zenith. This gradient was strongest at low solar altitudes.
(c) Introduction of the factor $[1 + \sin^3(TLT/2)]$ takes into account the effect of horizon brightening.
(d) Introduction of the factor $[(1+\cos^2 INC) \sin^3 Z]$ takes into account the effect of circumsolar radiation, also known as the sun's aureole, Z being the zenith angle of the sun.

Temps and Coulson's clear sky diffuse irradiance model may be expressed as

$$I_{D, TLT} = I_D \cos^2(TLT/2)[1 + \sin^3(TLT/2)](1 + \cos^2 INC) \sin^3 Z \tag{4.3.2.1}$$

4.3.2.2 Klucher's model

Klucher (1979) progressed the work of Temps and Coulson (1977) by developing an anisotropic model for all sky conditions. He used hourly measured radiation values for New York for a six-month period on surfaces tilted towards the equator at 37° and 60° angles. He found that the Liu–Jordan isotropic model gives good results under overcast skies, but underestimates insolation under clear and part-overcast conditions. Klucher noted that the Temps-Coulson model provides a good prediction for clear sky conditions but overestimates overcast insolation.

Klucher's results may be put explicitly as follows:

(a) The Liu–Jordan model provides a good fit for low intensities (< 300 W/m^2), such values being associated with overcast skies. However, for intensities greater than 500 W/m^2, the Liu–Jordan model underestimates insolation by as much as 80 W/m^2 at 37° tilt and up to 100 W/m^2 at 60° tilt.
(b) The Temps–Coulson model provided results which are highly consistent with the entire six months of data for clear skies. However, for other days, the model over-predicted insolation by about 120 W/m^2 in winter and up to 50 W/m^2 in summer.

Klucher's model is given by

$$I_{D,\text{TLT}} = I_D \cos^2(\text{TLT}/2) \, [1+F' \sin^3(\text{TLT}/2)] \, [1 + F' \cos^2 \text{INC} \sin^3 Z] \qquad (4.3.2.2)$$

where F' is the modulating function, given by Eq. (4.2.2). Under overcast skies F' becomes zero, thereby reducing Klucher's model to that of Liu and Jordan, and under clear skies F' approaches unity, thus approximating the Temps–Coulson model.

4.3.2.3 Hay's model
Hay's model (Hay, 1979; Hay and Davies, 1980; Hay and McKay, 1988) assumes diffuse radiation incident on a horizontal surface to be composed of circumsolar and uniform background sky-diffuse components. The two components are weighted accordingly, as given by

$$I_{D,\text{TLT}} = I_D \, [F \, r_B + (1 - F) \cos^2(\text{TLT}/2)] \qquad (4.3.2.3)$$

Ma and Iqbal (1983) have used a year's measured data for Woodbridge, Ontario to compare the performance of Klucher's and Hay's models. Two statistical measures of error, the root mean square error (RMSE) and the mean bias error (MBE), were used to evaluate the accuracy of the models. Ma and Iqbal have presented their results for measured and predicted global irradiation on south-facing slopes of 30°, 60° and 90°. Global slope irradiation was calculated as the sum of beam, diffuse and ground-reflected components; the last, in turn, was calculated using a daily value of measured ground-reflected irradiation. The conclusions of Ma and Iqbal may be listed as follows:

(a) The RMSEs for all three models increase with the slope.
(b) For all slopes, the models are reasonably accurate during the summer months.
(c) The errors are maximum in winter months when the sky-diffuse component forms a significant proportion of the total irradiation.
(d) The highest RMSE for the isotropic model is 30%. Hay and Klucher models are much alike with maximum RMSEs of 20% on 90° slopes and 15% on 60° slopes.

4.3.2.4 Skartveit and Olseth's model
This model was primarily developed for the higher latitudes (Skartveit and Olseth, 1986)

and may be expressed as

$$I_{D,TLT} = I_D [Fr_B + B \cos TLT + (1 - F - B) \cos^2(TLT/2)] \qquad (4.3.2.4)$$

where $B = \max[(0.3 - 2F), 0]$.

4.3.2.5 Reindl et al.'s model
Reindl et al. (1990) have used the work of Hay and Davies (1980) and Klucher (1979) to propose

$$I_{D,TLT} = I_D \{(1 - F) \cos^2(TLT/2) [1 + f \sin^3(TLT/2)] + Fr_B\} \qquad (4.3.2.5)$$

F is defined by Eq. (4.2.1) and $f = (I_B / I_G)^{0.5}$.

4.3.3 Third generation models

These models treat the sky-diffuse component as anisotropic. Most models under this category decompose non-overcast irradiance as the sum of two components, i.e. circumsolar and background sky-diffuse components, an exception being the Perez et al. (1990) model which consists of three components.

Three all-sky models which belong to this category, i.e. Gueymard (1987), Muneer (1987) and Perez et al. (1990), are described herein. The first two models owe their development to the works of Moon and Spencer (1942) and Steven and Unsworth (1979; 1980) which are presented first.

4.3.3.1 Moon and Spencer's model
Moon and Spencer (1942) used measured data to demonstrate a relationship between the luminance of a patch of an overcast sky and its zenith angle θ. Their proposed relationship was given as

$$L_\theta = L_z (1 + b \cos \theta) / (1 + b) \qquad (4.3.3.1)$$

The parameter b in this equation is the luminance (or radiance) distribution index. Moon and Spencer obtained the value of $b = 2$ which was later adopted to define the CIE standard overcast sky. Kondratyev (1969) has also reported that the radiance and luminance distributions are rather similar and azimuth independent for an overcast sky.

A further discussion on radiance and luminance distribution of overcast skies is provided in Section 4.5.

4.3.3.2 Steven and Unsworth's model
Steven and Unsworth (1980) measured the overcast sky radiance at Sutton Bonnington (52.8°N) by using nine pyrheliometers, eight of which faced north, east, south and west at angles of 30° and 60° to the zenith and the remaining one faced the zenith. Radiance distribution of the sky was found to be similar to Eq. (4.3.3.1) but they obtained the average value of $b = 1.23$. Steven and Unsworth attributed the difference in the value of

b, in the two studies, to the difference between radiance and luminance distribution of the overcast sky. Steven (1977a; 1977b) has also measured the radiance distribution of a clear sky and reported its anisotropic nature. Steven and Unsworth (1979) used Steven's (1977a; 1977b) measurements to predict irradiation on an inclined surface as given by

$$I_{D, TLT} = I_{DC}\, r_B + I_{DB} \tag{4.3.3.2}$$

where I_{DC} and I_{DB} are, respectively, the circumsolar and background-sky components of the hourly diffuse irradiation. I_{DC} was assumed to be proportional to I_D as given by

$$I_{DC} = s\, I_D \tag{4.3.3.3}$$

where s is a proportionality constant which varies with the sky clarity.

Steven (1977a) integrated Eq. (4.3.3.1), replacing luminance by radiance to obtain

$$\frac{I_{DB}}{I_D} = \cos^2\frac{TLT}{2} + \frac{2b}{\pi(3+2b)}\left[\sin TLT - TLT \cos TLT - \pi \sin^2\frac{TLT}{2}\right] \tag{4.3.3.4}$$

The mathematical proof for this equation is given below. The computation of hourly tilted surface irradiation thus requires I_D and the values of s and b. Steven and Unsworth (1979) have calculated these values for a range of zenith angles between 35° and 65°. For all zenith angles, taken together, they obtained $s = 0.51$ and $b = -0.87$.

Derivation of Eq. (4.3.3.4) for sky-diffuse irradiance on a tilted surface

The analysis, originally due to Steven (1977b), has been presented by Usher and Muneer (1989). Consider the solar geometry for a surface having a slope of angle α to the horizontal and normal $\hat{N}$ (Figure 4.3.1). Let $\hat{A}$ be a unit vector in the direction of the sun, θ be the zenith angle and ϕ be the azimuth angle. With respect to the Cartesian co-ordinates x, y, z in the horizontal plane,

$$\hat{A} = (\sin\theta\cos\phi,\ \sin\theta\sin\phi,\ \cos\theta)$$
$$\hat{N} = (\sin\alpha,\ 0,\ \cos\alpha)$$

If $I(\theta, \phi)$ denotes the radiance distribution then the flux of radiation on the tilted surface is given by

$$I_{D\alpha} = \iint_{\text{visible region of sky}} I(\theta,\phi)\,\hat{A}\,\hat{N}\, d\Omega$$

$\hat{N}$ is the vector drawn normal to the elemental sky patch $d\Omega$

$$I_{D\alpha} = \iint I(\theta,\phi)(\sin\theta\cos\phi\sin\alpha + \cos\theta\cos\alpha)\sin\theta\, d\theta\, d\phi \tag{4.3.3.5.}$$

This integral may be simplified as follows. First consider the projection of the unit

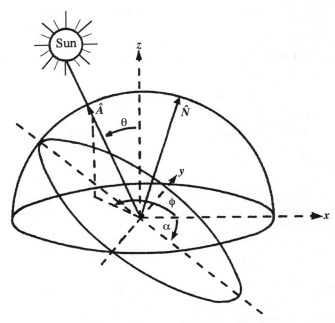

Figure 4.3.1 *Solar geometry for an inclined surface*

vector $\hat{A}$ onto the inclined surface, the projected sector **B** being given by

$$\mathbf{B} = \hat{A} - (\hat{A} \cdot \hat{N})\hat{N}$$

Now consider the transformation corresponding to the rotation of the inclined surface through an angle α about the y-axis which maps the vector **B** into a vector **C** in the horizontal xy-plane. The matrix of this transformation is given by

$$R_\alpha = \begin{pmatrix} \cos\alpha & 0 & -\sin\alpha \\ 0 & 1 & 0 \\ \sin\alpha & 0 & \cos\alpha \end{pmatrix}$$

$$\mathbf{C} = R_\alpha \mathbf{B}$$

It may then be shown that

$$\mathbf{C} = \begin{pmatrix} \cos\alpha\sin\theta\cos\phi - \sin\alpha\cos\theta \\ \sin\theta\sin\phi \\ 0 \end{pmatrix}$$

Thus, consider co-ordinates in the tilted plane given by

$$x' = \cos\alpha \sin\theta \cos\phi - \sin\alpha \cos\theta$$
$$y' = \sin\theta \sin\phi$$

With these co-ordinates Eq. (4.3.3.5) reduces to

$$I_{D\alpha} = \iint_{R'} I(\theta(x',y'))\phi(x',y'))\,dx'dy' \qquad (4.3.3.6)$$

where the region R' is the projection of the sky hemisphere on the tilted plane. For negative x', i.e. $\pi/2 \leq \phi \leq 3\pi/2$, the region R' is a semicircle of unit radius. However, for positive x', i.e. $-\pi/2 \leq \phi \leq \pi/2$, the region R' is bounded by an ellipse which is determined by the projection of the unit semicircle, $\pi/2 \leq \phi \leq \pi/2$, in the horizontal xy-plane, onto the sloped $x'y'$-plane. The parametric equation of this ellipse will be given by

$$x' = \cos\alpha \cos\phi, \quad -\pi/2 \leq \phi \leq \pi/2$$
$$y' = \sin\phi, \quad -\pi/2 \leq \phi \leq \pi/2$$

For subsequent evaluation of the integral in Eq. (4.3.3.6) it is convenient to transform to polar co-ordinates (r, η). Thus,

$$I_{D\alpha} = \iint_{R'} I(\theta(r,\eta))\phi(r,\eta))\,r\,dr\,d\eta \qquad (4.3.3.7)$$

A suitable expression for the radiance distribution $I(\theta, \phi)$ is now required. It may be assumed that:

(a) I is independent of ϕ, and
(b) I varies linearly with $\cos\theta$.

Thus, let $I(\theta) = I_{90}(1 + b\cos\theta)$, where I_{90} and b are constants to be determined. It may be shown that if the flux on a horizontal surface is I_D then

$$I(\theta) = \frac{I_D}{\pi(1 + 2b/3)}(1 + b\cos\theta) \qquad (4.3.3.8)$$

Using this expression for the radiance distribution and the appropriate polar co-ordinate transformation, Eq. (4.3.3.7) becomes

$$I_{D\alpha} = \iint_{R'} \left\{1 + b\left[-r\cos\eta\sin\alpha + \cos\alpha\sqrt{(1-r^2)}\right]\right\}\frac{I_D}{\pi(1+2b/3)}r\,dr\,d\eta$$

The limits of integration are found to be given by

$$0 \leq r \leq 1 \quad \text{for } \pi/2 \leq \eta \leq 3\pi/2$$
$$0 \leq r \leq q \quad \text{for } -\pi/2 \leq \eta \leq \pi/2$$

where

$$q = \cos \alpha / \sqrt{(1 - \sin^2 \alpha \sin^2 \eta)} \qquad (4.3.3.9)$$

Algebraic manipulation leads to

$$\frac{I_{D\alpha}}{I_D} = \cos^2\left(\frac{\alpha}{2}\right) + \frac{2b}{3\pi(1+2b/3)}\left[\sin\alpha - \alpha\cos\alpha - \pi\sin^2\left(\frac{\alpha}{2}\right)\right] \qquad (4.3.3.10)$$

The subscript α has been introduced to signify that this is the sky-diffuse irradiation on a surface inclined at an angle α to the horizontal. When $b = 0$, Eq. (4.2.2.10) reduces to the isotropic case.

Note that α used in the above derivation is more commonly known as the surface tilt angle TLT.

As mentioned above, Eq. (4.3.3.10) enables estimation of sky-diffuse irradiation on any tilted surface given the value of the radiance distribution index b. Steven and Unsworth (1979; 1980) have provided the value of b for the overcast and clear sky conditions. Their approach therefore falls short of an all-sky model. The above analysis has nevertheless been used effectively by Gueymard (1987) and Muneer (1987; 1990a; 1990b) to obtain an anisotropic model which enables computation under the varying sky conditions.

4.3.3.3 Gueymard's model

It was shown above that the sky-diffuse irradiance received on any surface is physically related to the radiance distribution. For a tilted receiving surface r_D may be defined as

$$r_D = I_{D,\text{TLT}} / I_D \qquad (4.3.3.11)$$

The main assumption used in the derivation of Gueymard's (1987) model is that the radiance of a partly cloudy sky may be considered as a weighted sum of the radiances of a clear and an overcast sky. Using the work of Steven and Unsworth (1979), Gueymard has introduced the concept of weighted normalised radiances which may be translated into an equation for the slope factor corresponding to general sky conditions,

$$r_D = (1 - N_{pt})\, r_{d0} + N_{pt}\, r_{d1} \qquad (4.3.3.12)$$

where N_{pt} is the weighing factor and subscripts 0 and 1 refer to the opacity (0 clear; 1 overcast). r_{d0} and r_{d1} may be obtained if the corresponding distributions of normalised radiances are available. r_{d0} is obtained as the sum of a circumsolar component (dependent on INC and SOLALT) and a hemispheric component (dependent on TLT and SOLALT) via polynomial regression:

$$r_{d0} = \exp(a_0 + a_1 \cos \text{INC} + a_2 \cos^2 \text{INC} + a_3 \cos^3 \text{INC}) + F(\text{TLT})\, G(\text{SOLALT})$$
(4.3.3.13)

where

$$a_0 = -0.897 - 3.364\, h' + 3.960\, h'^2 - 1.909\, h'^3 \qquad (4.3.3.14)$$

$$a_1 = 4.448 - 12.962\, h' + 34.601\, h'^2 - 48.784\, h'^3 + 27.511\, h'^4 \qquad (4.3.3.15)$$

$$a_2 = -2.770 + 9.164\, h' - 18.876\, h'^2 + 23.776\, h'^3 - 13.014\, h'^4 \qquad (4.3.3.16)$$

$$a_3 = 0.312 - 0.217\, h' - 0.805\, h'^2 + 0.318\, h'^3 \qquad (4.3.3.17)$$

$$F(\text{TLT}) = [1 + b_0 \sin^2(\text{TLT}) + b_1 \sin(2\text{TLT}) + b_2 \sin(4\text{TLT})]/[1 + b_0] \qquad (4.3.3.18)$$

$$G(\text{SOLALT}) = 0.408 - 0.323\, h' + 0.384\, h'^2 - 0.170\, h'^3 \qquad (4.3.3.19)$$

where $h' = 0.01$ SOLALT (degrees), $b_0 = -0.2249$, $b_1 = 0.1231$, $b_2 = -0.0342$.

Overcast sky condition

In this case $r_D = r_{d1}$ (refer to Eq. (4.3.3.12)). r_{d1} is the ratio of the short-wave energy received on the tilted surface to that incident on the horizontal. It is thus equal to the right hand side of Eq. (4.3.3.4). The limit value $b = 0$ in Eq. (4.3.3.4) would correspond to an isotropic sky. However, most theoretical and experimental determinations of b lie in the range 1.0–2.0. A mean value of $b = 1.5$ has been adopted by Gueymard.

Part-overcast sky condition

In the general case of a part-overcast sky, Eq. (4.3.3.12) applies, where r_{d0} is obtained from Eq. (4.3.3.13) and r_{d1} from the right hand side of Eq. (4.3.3.4). It was shown by Gueymard that b decreases with cloudiness. To take this effect into account b is assumed to be a linear function of the cloud opacity, such that $b = 0.5 + N_{pt}$.

If no cloud observation is available, but hourly possible sunshine (SF) is measured, the following relationship is suggested:

$$N_{pt} = 1 - \text{SF} \qquad (4.3.3.20)$$

If hourly sunshine is not available, N_{pt} is to be estimated from horizontal diffuse ratio thus:

$$N_{pt} = \max\{\min(Y, 1), 0\} \qquad (4.3.3.21)$$

where

$$Y = 6.6667\, (I_D/I_G) - 1.4167, \quad (I_D/I_G) \le 0.227 \qquad (4.3.3.22)$$

$$Y = 1.2121 \ (I_D/I_G) - 0.1758, \quad \text{otherwise} \tag{4.3.3.23}$$

4.3.3.4 Perez et al.'s model

The Perez et al. (1990) model is based on a three-component treatment of the sky-diffuse irradiance and illuminance. The incident diffuse energy on any tilted surface is given by

$$I_{D,\text{TLT}} = I_D \left[(1 - F_1) \cos^2(\text{TLT}/2) + F_1 (a_0/a_1) + F_2 \sin \text{TLT}\right] \tag{4.3.3.24}$$

where F_1 and F_2 are circumsolar and horizon brightness coefficients, and a_0 and a_1 are terms that account for the respective angles of incidence of circumsolar radiation on the tilted and horizontal surfaces (the circumsolar radiation is considered to be from a point source):

$$\begin{aligned} a_0 &= \max[0, \cos \text{INC}] \\ a_1 &= \max[\cos 85°, \cos z] \end{aligned} \tag{4.3.3.25}$$

With these definitions, a_0/a_1 becomes r_B.

The brightness coefficients F_1 and F_2 are functions of three parameters that describe the sky conditions: the zenith angle z, sky clearness ε_i, and a brightness Δ. ε is a function of I_D and the normal incidence beam irradiation $I_{B,n}$, i.e.

$$\varepsilon_i = \{[I_D + I_{B,n}/I_D] + 5.535 \times 10^{-6} z^3\} / \{1 + 5.535 \times 10^{-6} z^3\} \tag{4.3.3.26}$$

Table 4.3.1 Coefficients for Perez et al. (1990) slope irradiance and illuminance model (4.3.24)

ε (bin)	1	2	3	4	5	6	7	8
Lower bound	1.000	1.065	1.230	1.500	1.950	2.800	4.500	6.200
Upper bound	1.065	1.230	1.500	1.950	2.800	4.500	6.200	—

Irradiance

f_{11}	−0.0083	0.1299	0.3297	0.5682	0.8730	1.1326	1.0602	0.6777
f_{12}	0.5877	0.6826	0.4869	0.1875	−0.3920	−1.2367	−1.5999	−0.3273
f_{13}	−0.0621	−0.1514	−0.2211	−0.2951	−0.3616	−0.4118	−0.3589	−0.2504
f_{21}	−0.0596	−0.0189	0.0554	0.1089	0.2256	0.2878	0.2642	0.1561
f_{22}	0.0721	0.0660	−0.0640	−0.1519	−0.4620	−0.8230	−1.1272	−1.3765
f_{23}	−0.0220	−0.0289	−0.0261	−0.0140	0.0012	0.0559	0.1311	0.2506

Illuminance

f_{11}	0.0113	0.4296	0.8093	1.0141	1.2818	1.4257	1.4848	1.1695
f_{12}	0.5707	0.3634	−0.0535	−0.2522	−0.4205	−0.6533	−1.2139	−0.2998
f_{13}	−0.0820	−0.3066	−0.4422	−0.5311	−0.6888	−0.7789	−0.7837	−0.6149
f_{21}	−0.0947	0.0499	0.1809	0.2750	0.3802	0.4247	0.4111	0.5180
f_{22}	0.1579	0.0080	−0.1686	−0.3498	−0.5586	−0.7851	−0.6292	−1.8924
f_{23}	−0.0177	−0.0650	−0.0918	−0.0957	−0.1144	−0.0966	−0.0822	−0.0551

where z is in degrees, and

$$\Delta = m\, I_D / I_{E,n} \tag{4.3.3.27}$$

where m is the air mass and $I_{E,n}$ is the extraterrestrial normal incidence radiation. The brightness coefficients F_1 and F_2 are obtained via Table 4.3.1 and the following equations:

$$F_1 = \max[0, \{F_{11} + F_{12}\Delta + (\pi/180)\, z\, F_{13}\}] \tag{4.3.3.28}$$

$$F_2 = [F_{21} + F_{22}\Delta + (\pi/180)\, z\, F_{23}] \tag{4.3.3.29}$$

4.3.3.5 Muneer's model

Muneer's model (Muneer, 1987; 1990b; 1995; Saluja and Muneer, 1987) treats the shaded and sunlit surfaces separately and further distinguishes between overcast and non-overcast conditions of the sunlit surface. In this model the slope diffuse irradiation for surfaces in shade and sunlit surfaces under overcast sky is computed as

$$I_{D,TLT} = I_D\, [TF] \tag{4.3.3.30}$$

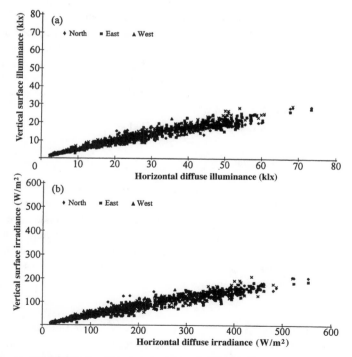

Figure 4.3.2 *Relationship between shaded vertical and horizontal diffuse (a) illuminance and (b) irradiance at Chofu*

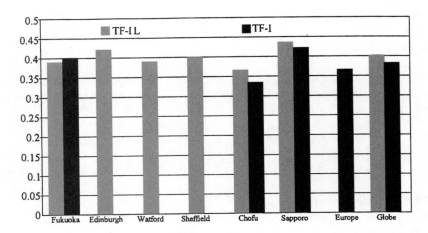

Figure 4.3.3 *Ratio of shaded vertical surface to horizontal diffuse incident energy*

Where TF_I is the tilt factor. TF_I represents the ratio of slope background diffuse irradiance to horizontal diffuse irradiance. Eq. (4.3.3.30) suggests that for any sloped surface in shade (facing away from the sun) there is a linear relationship between $I_{D, TLT}$ and I_D. Moreover, the value of the radiance (or luminance) distribution index b (refer to Eq. 4.3.3.31) may easily be obtained by finding a linear curve fit between a vertical surface irradiance and the horizontal diffuse irradiance. Figure 4.3.2 shows such a plot for Chofu in Japan. The scatter points shown in this figure represent 10-minute averaged data. The slopes of the best fit lines for Chofu as well as many other locations worldwide were obtained by Muneer (1987; 1990a; 1990b; 1995). Some of these best fit values are shown in Figure 4.3.3.

A sunlit surface under non-overcast sky is modelled as

$$I_{D, TLT} = I_D [TF_I(1 - F) + F\, r_B] \qquad (4.3.3.31)$$

TF_I is obtained from Eq. (4.3.3.4) using a value of b which corresponds to the appropriate sky and azimuthal condition. For the European climate a shaded surface is modelled with $b = 5.73$, while $b = 1.68$ for sun-facing surfaces under overcast sky (Muneer, 1990b). On a world-wide basis, Figure 4.3.3 suggests an average value of $b = 2.5$. Non-overcast skies, on the other hand, exhibit a continuously decreasing behaviour of b and therefore the following equations, obtained via data from 14 world-wide locations (Muneer, 1995) are recommended:

$$2b\, \{\pi\, (3 + 2b)\}^{-1} = 0.003\,33 - 0.415\, F - 0.6987\, F^2 \text{ for northern Europe} \qquad (4.3.3.32a)$$
$$2b\, \{\pi\, (3 + 2b)\}^{-1} = 0.002\,63 - 0.712\, F - 0.6883\, F^2 \text{ for southern Europe} \qquad (4.3.3.32b)$$
$$2b\, \{\pi\, (3 + 2b)\}^{-1} = 0.080\,00 - 1.050\, F - 2.8400\, F^2 \text{ for Japan} \qquad (4.3.3.32c)$$
$$2b\, \{\pi\, (3 + 2b)\}^{-1} = 0.040\,00 - 0.820\, F - 2.0260\, F^2 \text{ for the globe} \qquad (4.3.3.32d)$$

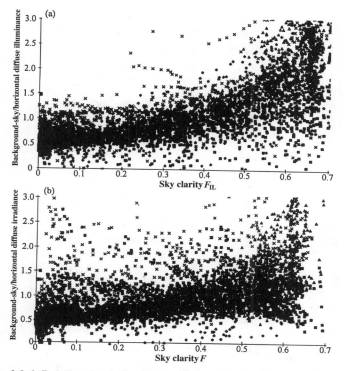

Figure 4.3.4 *Relationship between vertical sun-facing background-sky diffuse (a) illuminance and (b) irradiance fraction and sky clarity at Fukuoka*

A sample plot for Fukuoka, Japan which demonstrates the curvilinear relationship suggested by Eq. (4.3.3.32) is shown in Figure 4.3.4. Data from many other locations are presented in Figure. 4.3.5. These plots confirm the validity of the findings reported by Temps and Coulson (1977) and Steven and Unsworth (1979). Prog4-1.For enables slope irradiance computation using the procedure laid out for seven of the above models. These models are isotropic (first generation); Hay, Skartveit-Olseth, and Reindl et al. (second generation); and Gueymard, Perez et al., and Muneer (third generation).

Example 4.3.1

Using the isotropic sky-diffuse model, calculate the beam and diffuse irradiance on a vertical surface facing south in Edinburgh (55.95°N, 3.20°W) at 1120 hours on 10 August 1993. The 5-minute averaged values for horizontal global and diffuse irradiance, centred at 1120 hours, are respectively 552 and 267 W/m². Note: Ground-reflected radiation was excluded in these measurements.

For the above conditions,

$I_E = 980.15$ W/m²

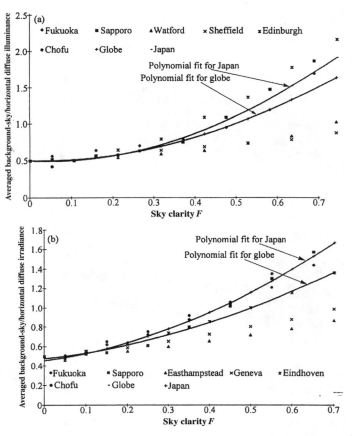

Figure 4.3.5 *Averaged background-sky diffuse (a) illuminance and (b) irradiance fraction versus sky clarity*

SOLALT = 48°. INC = 51.4°

TF = $\cos^2$ (TLT/2) = 0.5

r_B = cos INC/sin SOLALT = 0.8385

$I_{B,TLT}$ = $(G - D)\, r_B$ = (552 – 267) × 0.8385 = 239 W/m²

$I_{D,TLT} = I_D$ TF = 267 × 0.5 = 133.5 W/m²

I_{TLT} = 372.5 W/m²

The corresponding measured value is 379 W/m².

Example 4.3.2

Using Hay's model, perform calculations on the data given in Example 4.3.1.

We note from the preceding example that $I_E = 980.15$ W/m^2, TF = 0.5, $r_B = 0.8385$ and $I_{B,TLT} = 239$ W/m^2. Then

$$F = (I_G - I_D)/I_E = 0.291$$

$$I_{D,TLT} = I_D [Fr_B + (1 - F) \text{TF}] = 267 [0.291 \times 0.8385 + 0.709 \times 0.5] = 160 \text{ W/m}^2$$

$$I_{TLT} = 399 \text{ W/m}^2$$

The corresponding measured value is 379 W/m^2.

Example 4.3.3

Using Skartveit and Olseth's model, perform calculations on the data given in Example 4.3.1.

In this case, $B = \max[(0.3 - 2F), 0] = 0.0$, TF = 0.5 and $r_B = 0.8385$. Then

$$I_{D,TLT} = 267 [0.291 \times 0.8385 + 0.0 \times 0.0 + (1 - 0.291)0.5] = 160 \text{ W/m}^2$$

$$I_{TLT} = 399 \text{ W/m}^2$$

The corresponding measured value is 379 W/m^2.

Example 4.3.4

Using Reindl's model, perform calculations on the data given in Example 4.3.1.

We note that $F = 0.291, f = 0.719$. Then

$$I_{D,TLT} = 267 [(1 - 0.291) \times 0.5 \times (1 + 0.719 \times 0.3535) + 0.291 \times 0.8385] = 184 \text{ W/m}^2$$

$$I_{TLT} = 423 \text{ W/m}^2$$

The corresponding measured value is 379 W/m^2.

Example 4.3.5

Using Gueymard's model, perform calculations on the data given in Example 4.3.1.

We note that SOLALT = 48°, $h' = 0.48$. Using Equations (4.3.3.14)–(4.3.3.19),

126 SOLAR RADIATION AND DAYLIGHT MODELS

$a_0 = -1.81$, $a_1 = 2.26$, $a_2 = -0.78$, $a_3 = 0.0573$, $G(\text{SOLALT}) = 0.323$
since TLT = 90°, $F(\text{TLT}) = 1.00$

Then r_{d0} and r_{d1} are obtained (note INC = 51.4°) as

$r_{d0} = 0.8244$, $r_{d1} = 0.4314$

Now $N_{pt} = 0.4105$ as $Y = 0.4105$. Then Eq. (4.3.3.11) is used to obtain:

$I_{D,\text{TLT}} = 177 \text{ W/m}^2$

$I_{\text{TLT}} = 416 \text{ W/m}^2$

The corresponding measured value is 379 W/m².

Example 4.3.6

Using Perez et al.'s model, perform calculations on the data given in Example 4.3.1.

The data on I_B, I_E, cos INC, sin SOLALT and r_B from Example 4.3.2 are used here. Thus

$Z = 90 - \text{SOLALT} = 42°$
$m = 1.343$
$\varepsilon = 2.019$ Eq. (4.3.3.26)
$\Delta = 0.26231$ Eq. (4.3.3.27)

We note from Table 4.3.1 that the above value of ε lies in the fifth bin. Using the values of F_{11} to F_{23}, we obtain

$F_1 = 0.505\ 33$, $F_2 = 0.105\ 28$

$a = \max(0, \cos \text{INC}) = 0.6234$,
$b = \max(\cos 85°, \sin \text{SOLALT}) = 0.7435$

Then, using Eq. (4.3.3.24),

$I_{D,\text{TLT}} = I_D \left[(1 - F_1) \cos^2 (\text{TLT}/2) + F_1 (a_0/a_1) + F_2 \sin \text{TLT} \right] = 207 \text{ W/m}^2$

$I_{\text{TLT}} = 446 \text{ W/m}^2$

The corresponding measured value is 379 W/m²

Example 4.3.7

Using Muneer's model, perform calculations on the data given in Example 4.3.1.

First of all from Eqs (4.3.3.32) and (4.3.3.4) we find TF = 0.6. Then, using Eq. (4.3.3.31) and noting the values of F and $I_{B,TLT}$ from the above examples,

$$I_{D,TLT} = I_D [TF (1 - F) + Fr_B] = 139 \text{ W/m}^2$$

$$I_{TLT} = I_{D,TLT} + I_{B,TLT} = 418 \text{ W/m}^2$$

The corresponding measured value is 379 W/m².

The electronic data file File3-1.Csv, contained in the CD, provides 5 minute averages of horizontal and vertical irradiance and illuminance for Edinburgh for 11 and 12 August 1993 (respectively, an overcast and a clear day). Based on these data, Figures 4.3.6 and 4.3.7 have been prepared to enable performance evaluation of the above seven slope irradiance models. Further, Figures. 4.3.8–4.3.11 provide a finer examination of the models for a non-overcast day (12 August 1993). Tables 4.3.2 and 4.3.3 present the statistical evaluation of the models. The present analysis suggests that the accuracy of instantaneous predictions improves significantly when the isotropic model is replaced with most of the second generation and all of the third generation models. The shortcomings of most of the second generation models are noticeable in handling shaded surfaces (surfaces facing away from the sun), e.g. northern aspects in the northern hemisphere and vice versa. This is due to their assumption of isotropicity for such situations which is contrary to the measured evidence, as shown above.

Example 4.3.8
Long-term averaged measured horizontal and slope irradiation data are available for

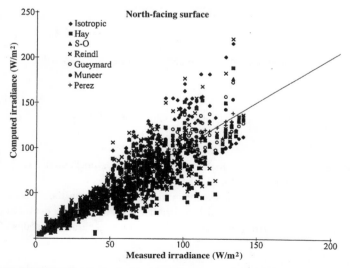

Figure 4.3.6 *Evaluation of slope irradiance models for a north-facing surface*

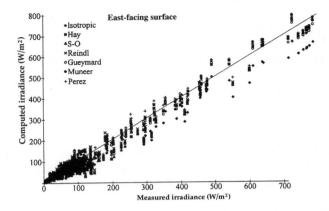

Figure 4.3.7 *Evaluation of slope irradiance models for an east-facing surface*

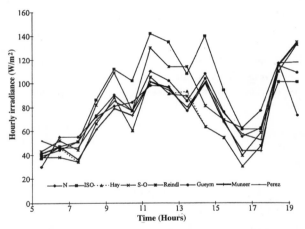

Figure 4.3.8 *Measured and estimated irradiance, north-facing surface*

Easthampstead, UK (Meteorological Office, 1980). Table 4.3.4 presents such data for June (top part).

(a) Using the daily totals, compute the hourly horizontal global and diffuse irradiation. Compare these with given hourly data.
(b) Compute the vertical irradiation for north and south using the above computed hourly horizontal global and diffuse irradiation. Perform the estimations using at least one model from each of the above three generation groups (Sections 4.3.1–4.3.3).
(c) Sum the computed hourly vertical surface irradiations to obtain corresponding daily totals.

Table 4.3.2 Evaluation of slope irradiance models at an hourly level, Edinburgh (55.95°N), August 1993 (W/m²)

		North	East	South
Isotropic	MBE	10	−5	10
	RMSE	24	36	25
Hay	MBE	−5	−8	−5
	RMSE	23	28	24
Skartveit and Olseth	MBE	−13	−16	−13
	RMSE	22	30	28
Reindl et al.	MBE	4	−1	4
	RMSE	24	25	24
Gueymard	MBE	−1	−7	−3
	RMSE	12	22	24
Muneer	MBE	−7	−6	1
	RMSE	13	22	27
Perez et al.	MBE	−2	−3	2
	RMSE	10	19	24

Table 4.3.3 Measured and computed slope irradiation for Edinburgh (55.95°N), 12 August 1993 (W/m²)

Hour	G	D	North	Iso	Hay	S–O	Reindl	Gueymard	Muneer	Perez
5.5	38	31	30	37	41	38	43	39	41	52
6.5	96	80	55	47	46	38	52	44	47	46
7.5	269	102	55	51	34	34	44	51	36	46
8.5	329	172	72	86	66	66	82	73	61	69
9.5	391	223	81	112	88	88	109	90	79	85
10.5	581	204	84	102	60	60	77	77	73	77
11.5	542	284	98	142	105	105	130	110	101	101
12.5	594	269	97	135	91	91	114	102	96	94
13.5	340	215	77	108	93	89	114	85	77	80
14.5	763	280	104	140	63	63	81	108	100	101
15.5	507	187	69	94	54	54	69	76	67	75
16.5	422	121	62	61	30	30	39	55	43	57
17.5	221	122	77	61	47	47	58	62	43	52
18.5	132	64	117	101	110	110	116	115	115	117
19.5	32	22	73	101	133	133	135	109	134	118
Hour	G	D	East	Iso	Hay	S–O	Reindl	Gueymard	Muneer	Perez
5.5	38	31	83	80	95	92	97	90	95	129
6.5	96	80	170	104	118	110	123	115	118	146
7.5	269	102	542	454	518	518	528	519	527	540
8.5	329	172	414	336	380	380	396	392	390	404

Continued overleaf

SOLAR RADIATION AND DAYLIGHT MODELS

Table 4.3.3 *Continued*

Hour	G	D	East	Iso	Hay	S–O	Reindl	Gueymard	Muneer	Perez
9.5	391	223	332	288	313	313	334	325	325	343
10.5	581	204	376	339	350	350	367	371	370	379
11.5	542	284	210	212	194	194	220	213	213	201
12.5	594	269	137	135	91	91	114	122	96	94
13.5	340	215	91	108	93	89	114	89	77	80
14.5	763	280	106	140	63	63	81	108	100	101
15.5	507	187	76	94	54	54	69	70	67	75
16.5	422	121	55	61	30	30	39	45	43	57
17.5	221	122	44	61	47	47	58	48	43	52
18.5	132	64	29	32	23	23	29	26	23	30
19.5	32	22	8	11	9	9	11	9	8	12

Hour	G	D	South	Iso	Hay	S–O	Reindl	Gueymard	Muneer	Perez
5.5	38	31	13	16	15	12	17	13	11	15
6.5	96	80	44	40	38	30	43	35	28	34
7.5	269	102	131	96	89	89	98	106	97	102
8.5	329	172	202	176	179	179	195	192	189	190
9.5	391	223	271	235	245	245	266	256	257	265
10.5	581	204	467	411	438	438	455	456	457	469
11.5	542	284	400	364	391	391	417	407	410	422
12.5	594	269	464	418	450	450	474	468	472	497
13.5	340	215	250	214	224	219	244	233	230	257
14.5	763	280	524	523	568	568	585	580	599	583
15.5	507	187	322	312	326	326	342	344	345	349
16.5	422	121	209	205	204	204	213	224	217	227
17.5	221	122	73	68	56	56	67	71	63	61
18.5	132	64	34	32	23	23	29	31	23	30
19.5	32	22	9	11	9	9	11	9	8	12

Hour	G	D	West	Iso	Hay	S–O	Reindl	Gueymard	Muneer	Perez
5.5	38	31	11	16	15	12	17	13	11	15
6.5	96	80	31	40	38	30	43	33	28	34
7.5	269	102	44	51	34	34	44	39	36	46
8.5	329	172	70	86	66	66	82	67	61	69
9.5	391	223	91	112	88	88	109	87	79	85
10.5	581	204	89	102	60	60	77	80	73	77
11.5	542	284	124	142	105	105	130	121	101	101
12.5	594	269	148	157	119	119	142	151	140	126
13.5	340	215	182	159	157	152	177	161	163	166
14.5	763	280	554	522	567	567	585	580	599	583
15.5	507	187	550	494	554	554	569	562	572	577
16.5	422	121	716	628	711	711	720	703	724	725
17.5	221	122	448	350	419	419	430	411	426	431
18.5	132	64	454	391	478	478	483	446	482	482
19.5	32	22	170	222	300	300	302	239	301	261

Iso: Isotropic
S–O: Skartveit and Olseth.

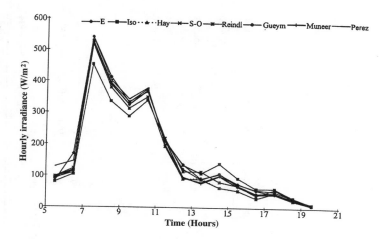

Figure 4.3.9 Measured and estimated irradiance, east-facing surface

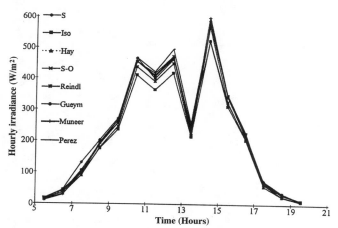

Figure 4.3.10 Measured and estimated irradiance, south-facing surface

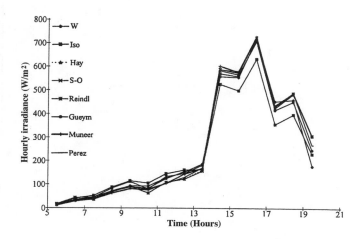

Figure 4.3.11 Measured and estimated irradiance, west-facing surface

Table 4.3.4 Evaluation of r_G Eq. (3.1.1), r_D Eq. (3.2.1), Isotropic Eq. 4.3.1.1), Reindl et al. Eq. (4.3.2.5) and Muneer Eq. 4.3.3.31) models

		Hour													Daily totals	Error (%)
	5.5	6.5	7.5	8.5	9.5	10.5	11.5	12.5	13.5	14.5	15.5	16.5	17.5	18.5		
Measured data for June																
G	108	214	325	411	483	533	547	539	514	467	392	306	208	111	5224	
D	67	117	164	206	239	269	283	281	267	244	214	169	122	72	2767	
North	100	100	83	86	92	97	100	100	97	92	86	81	97	97	1405	
South	31	53	94	164	231	269	283	275	253	208	150	89	53	31	2205	
Computed values																
G	116	203	299	393	476	538	571	571	538	476	393	299	203	116	5192	
D	74	122	167	208	242	265	278	278	265	242	208	167	122	74	2712	
North																
Isotropic	98	93	84	104	121	133	139	139	133	121	104	84	93	98	1541	9.7
Reindl et al.	115	102	83	102	116	125	130	130	125	116	102	83	102	115	1544	9.9
Muneer	110	95	60	74	86	94	99	99	94	86	74	60	95	110	1237	−12.0
South																
Isotropic	37	61	92	159	221	269	296	296	269	221	159	92	61	37	2268	2.8
Reindl et al.	39	63	93	170	240	295	325	325	295	240	170	93	63	39	2448	11.0
Muneer	26	44	85	161	231	286	316	316	286	231	161	85	44	26	2297	4.1

Comment on the accuracy obtained via the above calculation chain, e.g. horizontal daily total to horizontal hourly irradiation, hourly slope calculation and then summing to obtain daily slope irradiation.

Table 4.3.4 shows the results for the computational chain, e.g. from horizontal daily to horizontal hourly to slope hourly values. The comparison of daily totals for slope irradiation shows a good performance for all three models: isotropic, Reindl et al. and Muneer. Note that the isotropic model is on a par with the other two models considered herein. This is due to the considerable dampening effect in the averaging of data over very many days. However, on an instantaneous or an hour-by-hour level, the isotropic model does generate the order of errors shown above, e.g. Figures 4.3.8–4.3.11 and Table 4.3.3.

4.4 Slope illuminance models

In Chapter 3 models were presented which enable horizontal diffuse and global illuminance to be estimated from the corresponding irradiance quantities. The next stage in the analysis is to predict slope illuminance. Slope illuminance models may be used effectively to characterise the dynamic variations of the internal illuminance in buildings. The building simulation package SERI-RES (Haves, 1988) uses such a scheme.

The concept of the CIE overcast sky daylight factor was introduced in Chapter 3. It was shown in Section 4.3 that under non-overcast conditions, the sky is brighter in the vicinity of the sun and the horizon. Orientation factors, such as those given in *CIBSE Guide A2* (CIBSE, 1982), may be used to take account of windows facing the sun. However, orientation factors fail to take account of the dynamic variations in daylight, as pointed out by Haves and Littlefair (1988). The latter team have introduced a procedure to address the above mentioned shortcomings of the CIE daylight factor. Haves and Littlefair's (1988) procedure to obtain internal illuminance is summarised as follows:

$$IL_{in} = IL_{TLT} \, r \tag{4.4.1}$$

where IL_{in} is the desired horizontal internal illuminance and IL_{TLT} the external slope illuminance. r is defined as the modified daylight factor, it being the ratio of the internal to external illuminance in the plane of the window with a tilt angle of TLT:

$$r = d \, / \, d_{TLT} \tag{4.4.2}$$

where d is the standard CIE daylight factor (see Chapter 3) and d_{TLT} is the ratio of the exterior illuminance in the plane of the window to the exterior horizontal illuminance under CIE overcast sky conditions. For an unshaded vertical window, $d_{TLT} = 0.396 + 0.5\rho$. The procedure for obtaining ρ, the albedo of the underlying land mass, is presented in Chapter 6. Two models which enable estimation of IL_{TLT} are presented below.

4.4.1 Perez et al.'s model

The Perez et al. (1990) illuminance model is based on the same principles as its irradiance counterpart. Thus, Eqs. (4.3.3.24)–(4.3.3.29) are applicable with IL_G and IL_D respectively substituted for I_G and I_D. The required coefficients are given in Table 4.3.1.

4.4.2 Muneer and Angus's model

This model (Muneer and Angus, 1993; Muneer, 1995) is along the lines of Muneer's (1990b) work on solar irradiance modelling. Thus, the following equations are analogous to the corresponding slope irradiance relationships (see Section 4.3.3.5):

$$IL_\beta = (IL_G - IL_D)r_B + IL_D[(1 - F_{IL})TF + F_{IL}r_B] \qquad (4.4.3)$$

where

$$F_{IL} = (IL_G - IL_D)/IL_E \qquad (4.4.4)$$

TF_{IL} is obtained from Eq. (4.3.3.4) after estimating b as follows:

$$\frac{2b}{\{\pi(3 + 2b)\}} = 0.111 F_{IL} - 4.317 F_{IL}^2 \qquad (4.4.5)$$

IL_E in Eq. (4.4.4) is the horizontal extraterrestrial illuminance, the product of extraterrestrial irradiance and its luminous efficacy of 96.8 lm/W (Muneer and Angus, 1993).

Prog4-2.For enables computation of slope illuminance based on the Perez et al. model and the Muneer and Angus model.

Example 4.4.1

Table 4.4.1 provides the measured horizontal and slope illuminance data for Edinburgh (55.95°N, 3.2°W). Obtain vertical surface illuminance for the principal aspects and compare these with the above measured values.

The computed values of slope illuminance, obtained via Prog4-2.For, have been included in Table 4.4.1 to enable the required comparison. The MBEs show a remarkable performance by both models. As a matter of fact the errors involved with the above mentioned third generation irradiance and illuminance models are of the order of the measurement uncertainties, as also pointed out by Perez et al. (1990). A fuller discussion on the measurement errors is provided in Chapter 1.

Table 4.4.1 Comparison of measured and computed illuminance for Edinburgh, 12 August 1993

Hour	I_G (W/m²)	I_D (W/m²)	IL_G klx	IL_D klx	IL_N klx	IL_E klx	IL_S klx	IL_W klx	North Muneer	North Perez	East Muneer	East Perez	South Muneer	South Perez	West Muneer	West Perez
8.5	329	172	37.9	22.0	10.0	46.3	25.7	9.2	7.8	8.6	44.4	47.9	22.7	22.8	7.8	8.6
9.5	391	223	44.8	28.0	11.1	38.1	34.3	11.8	10.0	10.2	37.1	41.5	29.9	32.0	10.0	10.2
10.5	581	204	66.5	27.0	11.7	43.8	59.1	12.1	9.6	9.4	46.8	45.2	56.6	55.9	9.6	9.4
11.5	542	284	62.1	35.9	13.6	25.0	50.7	17.0	12.8	11.9	27.2	24.3	48.5	51.5	12.8	11.9
12.5	594	269	67.6	34.4	13.1	16.8	58.9	20.1	12.3	10.8	12.3	10.8	56.6	60.0	20.5	14.6
13.5	340	215	37.9	26.7	10.7	11.4	31.3	23.0	9.5	9.3	9.5	9.3	24.7	31.5	18.3	20.0
14.5	763	280	85.5	37.3	14.1	13.5	66.4	69.6	13.3	12.2	13.3	12.2	75.2	67.3	75.2	67.2
15.5	507	187	57.0	24.7	10.2	9.6	41.4	67.8	8.8	9.3	8.8	9.3	42.6	40.2	67.2	65.9
16.5	422	121	48.3	16.4	9.5	7.6	28.8	86.3	5.8	7.3	5.8	7.3	28.6	26.1	85.9	81.2
								MBE	−1.6	−1.7	−0.8	−0.5	−1.2	−1.0	−1.1	−3.1

4.5 Radiance and luminance distributions

The diffuse irradiance of sloping surfaces may be either measured, or calculated from the angular sky radiance distribution. Direct measurements, however, often include a component of radiation reflected onto the slope from neighbouring surfaces. This reflected radiation can be a large fraction of the diffuse irradiance, particularly on steep slopes and under cloudless skies when the ratio of diffuse to global radiation is small. On the other hand, separation of direct and diffuse solar radiation components by instruments with shade rings or disks excludes a fraction of circumsolar radiation which depends on shade ring dimensions. The alternative approach of calculating the diffuse irradiance of a sloping plane by integrating the radiance distribution of the sky 'seen' by the plane is attractive.

On the daylighting front, recent advances in computer graphics technology allow the realistic modelling of complex building interiors with a minimal training time. One such modelling and visualisation package is the RADIANCE lighting simulation system (Ward, 1994). The main features of RADIANCE have been enumerated by Mardaljevic (1995) as follows:

(a) It is a physically based program which allows precise estimation of interior illuminance.
(b) It has the capability to model geometries with realistic luminance distributions.
(c) It supports a wide variety of reflection and transmission models.
(d) It can import building scene descriptions from CAD systems.
(e) It can estimate daylight factors using real sky luminance distributions, rather than assuming the worst case scenario used by the CIE overcast sky model.

Mardaljevic (1995) has undertaken the first comparison of RADIANCE results with actual room illuminances under a real sky. Using 700 scans of measured sky luminance measurements, undertaken by the Building Research Establishment at Watford, illuminance predictions were found to agree quite favourably with the internal illuminances.

The RADIANCE system is free, copyrighted UNIX-based software produced at the Lawrence Berkeley Laboratory in California. Details of acquiring this software have been given by Mardaljevic (1995).

With the ever increasing processing power of PCs, more development and simulation tools such as RADIANCE will be used by architectural practices and building services engineers. The main engine in packages such as RADIANCE is the sky luminance distribution model. A discussion of some of these models follows.

4.5.1 Overcast sky distributions

Several authors have described the radiance distribution of overcast skies by an equation similar in form to Eq. (4.3.3.1),

$$R_\theta = R_z (1+b \cos \theta) / (1 + b) \tag{4.5.1}$$

where R_θ and R_z are expressed in W/Sr m². R_θ and R_z are the respective radiance counterparts of L_θ and L_z.

The number $1 + b$ is the ratio of radiance at the zenith to that at the horizon. The standard overcast sky (Moon and Spencer (1942)) uses this formula with $b = 2$ but Walsh (1961) suggested that $b = 1.5$ fitted the mean overcast sky more accurately. Goudriaan (1977) and Fritz (1955) gave a theoretical foundation for this form of distribution on the basis of an analysis of scattering and attenuation in clouds. Goudriaan's analysis shows that the value of b depends on surface albedo ρ according to $b = 2(1 - \rho)/(1 + 2\rho)$. For a typical range of surface albedo of 0.1 to 0.2, b should vary between 1.5 and 1.14. Fritz derived the relationship of b with ρ as $b = 1.5(1 - \rho)/(1 + \rho)$. With the same range of surface albedo, 0.1 to 0.2, the value of b according to the Fritz formula varies between 1.23 and 1.0. Table 4.5.1 summarises the results obtained by the above authors.

Table 4.5.1 Values of coefficient b in Eq. (4.5.1), overcast sky

Isotropic sky	0
Moon and Spencer (1942)	2
Fritz (1955)	1.0 to 1.23
Walsh (1961)	1.5
Goudriaan (1977)	1.14 to 1.5
Steven and Unsworth (1980)	1.12 to 1.36
Gueymard (1987)	1.5
Muneer (1987)	1.68

The angular distribution of overcast radiation was measured in detail by Steven and Unsworth (1980). Their averaged $b = 1.23$ is smaller than Moon and Spencer's value given above.

Under overcast skies Grace (1971) has noted rapid changes of radiance distribution when instantaneous measurements are carried out. However, hourly averages are less variable, e.g. Steven and Unsworth (1980) have shown that the coefficient of variation for b is typically about 0.2.

Nagata (1990a) has undertaken a series of radiance distribution measurements at Fukui, Japan. Based on several years of data, he has proposed the following equation for overcast sky radiance distribution:

$$R_\theta = R_z (2 + 3 \sin \theta) / 5 \qquad (4.5.2)$$

Integrating the above distribution results in

$$I_G = (4\pi/5) R_z \qquad (4.5.3)$$

from which R_z may be obtained if the value of I_G is available. Nagata (1990b) has also provided clear sky radiance distribution functions. These functions are similar in nature to those proposed by Steven and Unsworth (1979). The clear sky radiance distributions are given below.

4.5.2 Clear sky distributions

Measurements of the photometric or luminance distribution of a clear sky have been reported by Kimball and Hand (1921), Peyre (1927), Hopkinson (1954), Dogniaux (1954) and Kondratyev (1969). Most of these published results represent only one or two positions of the sun in the sky on a limited number of occasions. However, the distribution of diffuse radiation is strongly dependent on the solar zenith distance and varies to some extent with atmospheric turbidity.

The distribution of clear sky diffuse radiation was first explained theoretically by Lord Rayleigh (1871). Since then a number of attempts have been made to fit the theory to the observed distributions. In Pokrowski's (1929) formulation, R is expressed as a function of zenith angle θ and azimuth ϕ of a sky patch and ξ, the angle between the point (θ, ϕ) and the sun:

$$R(\theta, \xi) = S\{(1 + \cos^2\xi)/(1 - \cos\xi) + k\}\{1 - \exp(-\rho_s \sec \theta)\} \quad (4.5.4)$$

where S is a scaling factor, ρ_s is a scattering coefficient, and k is an empirical constant to allow for multiple scattering. Pokrowski proposed the values 0.32 and 5 for ρ_s and k respectively. Hopkinson (1954), however, found better agreement with measured luminance distributions with the arbitrary constant omitted, i.e. with $k = 0$ and $\rho_s = 0.32$.

Steven (1977a) measured the radiance from cloudless skies at Sutton Bonnington on a large number of days over a wide range of turbidities. When the measurements of radiance were normalised with respect to the diffuse irradiance of a horizontal surface, they were found to be independent of turbidity. On this basis standard distributions of radiance from cloudless skies were proposed as

$$R(\theta, \xi) = \pi^{-1}\{d_1 + d_2 \exp(d_3\xi) + d_4 \cos2\xi\}\{1 - \exp(d_5 \sec \theta)\} \quad (4.5.5)$$

Table 4.5.2 provides the coefficients for the above model.

4.5.3 Intermediate sky distributions

The CIE overcast sky estimates the minimum luminance daylight quantity, as it

Table 4.5.2 Coefficients to be used in Eq. (4.5.5)

	Solar zenith angle, (degrees)			
	35	45	55	65
d_1	0.61	0.65	0.73	0.76
d_2	11.90	10.70	11.10	13.00
d_3	−2.97	−2.82	−2.97	−3.09
d_4	−0.12	−0.20	−0.07	−0.17
d_5	−0.45	−0.48	−0.48	−0.42

Table 4.5.3 Measured luminance distribution data for intermediate sky, SOLALT=10 degrees (Nakamura et al., 1985)

AS (degrees)	Altitude angle of the sky patch (degrees)									
	0	10	20	30	40	50	60	70	80	90
0	2.82	3.44	3.81	3.60	2.80	1.97	1.50	1.26	1.10	1.00
10	2.68	3.20	3.54	3.33	2.62	1.95	1.50	1.25	1.09	1.00
20	2.42	2.80	3.04	2.93	2.43	1.91	1.50	1.25	1.08	1.00
30	2.13	2.46	2.73	2.60	2.20	1.81	1.47	1.24	1.08	1.00
40	1.91	2.19	2.38	2.30	2.03	1.75	1.45	1.22	1.08	1.00
50	1.79	1.98	2.10	2.06	1.90	1.68	1.42	1.21	1.07	1.00
60	1.70	1.80	1.89	1.90	1.80	1.59	1.39	1.19	1.07	1.00
70	1.60	1.71	1.79	1.80	1.70	1.52	1.34	1.16	1.06	1.00
80	1.40	1.60	1.71	1.72	1.62	1.44	1.27	1.13	1.05	1.00
90	1.20	1.43	1.60	1.63	1.52	1.34	1.20	1.10	1.03	1.00
100	1.06	1.25	1.43	1.46	1.38	1.23	1.11	1.04	1.01	1.00
110	0.97	1.09	1.20	1.24	1.22	1.10	1.01	0.98	0.98	1.00
120	0.95	1.00	1.06	1.06	1.01	0.96	0.94	0.94	0.96	1.00
130	0.94	0.94	0.95	0.94	0.91	0.92	0.91	0.92	0.95	1.00
140	0.93	0.92	0.90	0.90	0.91	0.91	0.90	0.91	0.95	1.00
150	0.92	0.90	0.91	0.94	0.94	0.92	0.90	0.91	0.94	1.00
160	0.90	0.92	0.98	1.01	0.99	0.94	0.91	0.90	0.94	1.00
170	0.93	0.96	1.01	1.05	1.04	0.98	0.92	0.90	0.94	1.00
180	0.94	0.97	1.03	1.07	1.05	1.00	0.93	0.90	0.93	1.00

AS = Azimuthal separation between the sky patch and sun.

Table 4.5.4 Measured luminance distribution data for intermediate sky, SOLALT=20 degrees (Nakamura et al., 1985)

AS (degrees)	Altitude angle of the sky patch (degrees)									
	0	10	20	30	40	50	60	70	80	90
0	2.53	5.00	14.00	5.10	3.49	2.51	1.87	1.42	1.17	1.00
10	2.37	4.26	5.23	4.50	3.41	2.46	1.83	1.40	1.16	1.00
20	2.08	3.28	3.81	3.72	3.10	2.37	1.75	1.38	1.15	1.00
30	1.83	2.53	3.00	3.00	2.70	2.15	1.67	1.35	1.14	1.00
40	1.65	1.92	2.30	2.49	2.34	1.95	1.60	1.33	1.13	1.00
50	1.53	1.73	1.95	2.08	2.02	1.80	1.54	1.31	1.11	1.00
60	1.42	1.61	1.73	1.83	1.80	1.68	1.46	1.25	1.10	1.00
70	1.29	1.52	1.65	1.70	1.64	1.52	1.34	1.18	1.08	1.00
80	1.08	1.35	1.53	1.58	1.52	1.37	1.21	1.10	1.03	1.00
90	0.98	1.16	1.33	1.40	1.33	1.21	1.10	1.03	1.01	1.00
100	0.94	1.00	1.11	1.20	1.16	1.07	1.00	0.98	0.99	1.00
110	0.90	0.92	0.98	1.02	1.00	0.95	0.93	0.94	0.96	1.00
120	0.86	0.88	0.90	0.89	0.88	0.88	0.88	0.90	0.94	1.00
130	0.84	0.84	0.83	0.81	0.81	0.82	0.83	0.86	0.92	1.00
140	0.83	0.83	0.81	0.80	0.80	0.80	0.81	0.84	0.90	1.00
150	0.83	0.82	0.80	0.81	0.81	0.81	0.81	0.83	0.90	1.00
160	0.82	0.81	0.80	0.81	0.81	0.81	0.80	0.82	0.89	1.00
170	0.82	0.81	0.81	0.82	0.82	0.81	0.80	0.82	0.89	1.00
180	0.81	0.80	0.82	0.83	0.82	0.81	0.80	0.82	0.89	1.00

AS = Azimuthal separation between the sky patch and sun.

Table 4.5.5 Measured luminance distribution data for intermediate sky, SOLALT=30 degrees (Nakamura et al., 1985)

AS degrees	Altitude angle of the sky patch (degrees)									
	0	10	20	30	40	50	60	70	80	90
0	2.09	3.00	5.20	11.50	5.08	2.93	2.04	1.56	1.25	1.00
10	2.00	2.72	4.20	5.20	4.12	2.86	2.03	1.55	1.24	1.00
20	1.70	2.24	2.86	3.25	2.95	2.48	1.96	1.53	1.23	1.00
30	1.52	1.86	2.23	2.56	2.46	2.17	1.86	1.49	1.12	1.00
40	1.40	1.66	1.89	2.15	2.19	2.02	1.73	1.41	1.18	1.00
50	1.30	1.53	1.71	1.89	1.94	1.78	1.53	1.29	1.22	1.00
60	1.19	1.38	1.55	1.66	1.68	1.56	1.34	1.19	1.08	1.00
70	1.03	1.23	1.39	1.46	1.47	1.38	1.21	1.11	1.05	1.00
80	0.94	1.09	1.21	1.30	1.28	1.20	1.13	1.07	1.03	1.00
90	0.87	0.98	1.08	1.15	1.11	1.06	1.04	1.02	1.00	1.00
100	0.82	0.90	0.95	0.97	0.96	0.95	0.94	0.95	0.97	1.00
110	0.78	0.82	0.83	0.83	0.84	0.85	0.86	0.90	0.95	1.00
120	0.75	0.75	0.75	0.74	0.75	0.77	0.80	0.85	0.92	1.00
130	0.73	0.71	0.70	0.69	0.70	0.72	0.75	0.81	0.89	1.00
140	0.70	0.69	0.68	0.68	0.68	0.69	0.72	0.79	0.87	1.00
150	0.68	0.67	0.67	0.67	0.68	0.68	0.71	0.77	0.86	1.00
160	0.66	0.66	0.66	0.67	0.68	0.67	0.70	0.76	0.85	1.00
170	0.64	0.65	0.66	0.68	0.68	0.67	0.69	0.76	0.85	1.00
180	0.63	0.65	0.67	0.68	0.68	0.67	0.69	0.75	0.84	1.00

AS = Azimuthal separation between the sky patch and sun.

Table 4.5.6 Measured luminance distribution data for intermediate sky, SOLALT=40 degrees (Nakamura et al., 1985)

AS degrees	Altitude angle of the sky patch (degrees)									
	0	10	20	30	40	50	60	70	80	90
0	1.53	1.87	2.26	3.30	7.60	3.50	2.26	1.63	1.29	1.00
10	1.50	1.82	2.16	2.94	3.95	3.08	2.19	1.62	1.27	1.00
20	1.40	1.72	1.95	2.40	2.65	2.45	2.04	1.57	1.24	1.00
30	1.30	1.57	1.72	1.93	2.60	2.03	1.80	1.48	1.21	1.00
40	1.19	1.42	1.59	1.74	1.79	1.75	1.59	1.38	1.18	1.00
50	1.08	1.28	1.43	1.54	1.60	1.53	1.42	1.28	1.13	1.00
60	0.96	1.16	1.30	1.37	1.42	1.36	1.27	1.18	1.08	1.00
70	0.85	1.03	1.16	1.26	1.27	1.22	1.16	1.10	1.05	1.00
80	0.75	0.91	1.04	1.13	1.14	1.12	1.09	1.06	1.03	1.00
90	0.64	0.80	0.92	1.01	1.03	1.02	1.02	1.01	1.00	1.00
100	0.57	0.71	0.80	0.86	0.89	0.90	0.91	0.94	0.97	1.00
110	0.53	0.62	0.69	0.74	0.77	0.79	0.83	0.88	0.94	1.00
120	0.50	0.57	0.62	0.65	0.68	0.71	0.76	0.82	0.91	1.00
130	0.48	0.53	0.57	0.60	0.63	0.66	0.70	0.77	0.87	1.00
140	0.46	0.51	0.54	0.56	0.59	0.62	0.67	0.74	0.85	1.00
150	0.44	0.49	0.52	0.54	0.56	0.59	0.64	0.72	0.83	1.00
160	0.43	0.48	0.51	0.52	0.54	0.57	0.62	0.70	0.82	1.00
170	0.42	0.47	0.50	0.51	0.53	0.56	0.61	0.69	0.81	1.00
180	0.42	0.47	0.49	0.51	0.52	0.55	0.60	0.68	0.80	1.00

AS = Azimuthal separation between the sky patch and sun.

Table 4.5.7 Measured luminance distribution data for intermediate sky, SOLALT=50 degrees (Nakamura et al., 1985)

AS degrees	Altitude angle of the sky patch (degrees)									
	0	10	20	30	40	50	60	70	80	90
0	1.11	1.47	1.67	2.03	2.82	5.60	3.05	2.02	1.39	1.00
10	1.10	1.44	1.67	2.00	2.60	3.45	2.80	1.95	1.35	1.00
20	1.05	1.36	1.58	1.87	2.25	2.59	2.35	1.85	1.34	1.00
30	1.00	1.22	1.45	1.67	1.93	2.06	1.99	1.67	1.31	1.00
40	0.95	1.07	1.23	1.43	1.62	1.72	1.66	1.50	1.25	1.00
50	0.90	0.98	1.06	1.21	1.34	1.43	1.42	1.32	1.18	1.00
60	0.84	0.92	0.97	1.06	1.15	1.22	1.23	1.18	1.08	1.00
70	0.75	0.84	0.87	0.92	1.00	1.07	1.10	1.08	1.04	1.00
80	0.65	0.73	0.78	0.81	0.86	0.92	0.99	1.02	1.03	1.00
90	0.53	0.62	0.68	0.72	0.77	0.82	0.89	0.96	0.99	1.00
100	0.47	0.55	0.61	0.64	0.68	0.72	0.80	0.90	0.96	1.00
110	0.42	0.50	0.55	0.58	0.61	0.65	0.72	0.82	0.92	1.00
120	0.38	0.46	0.51	0.53	0.56	0.60	0.67	0.75	0.88	1.00
130	0.36	0.43	0.47	0.49	0.52	0.56	0.62	0.70	0.84	1.00
140	0.35	0.41	0.45	0.47	0.49	0.52	0.58	0.66	0.81	1.00
150	0.34	0.40	0.44	0.45	0.47	0.50	0.55	0.64	0.78	1.00
160	0.34	0.39	0.43	0.44	0.46	0.49	0.54	0.62	0.77	1.00
170	0.33	0.39	0.42	0.43	0.45	0.48	0.53	0.62	0.76	1.00
180	0.33	0.39	0.42	0.43	0.44	0.47	0.53	0.61	0.76	1.00

AS = Azimuthal separation between the sky patch and sun.

Table 4.5.8 Measured luminance distribution data for intermediate sky, SOLALT=60 degrees (Nakamura et al., 1985)

AS degrees	Altitude angle of the sky patch (degrees)									
	0	10	20	30	40	50	60	70	80	90
0	0.53	0.67	0.82	1.08	1.57	2.55	4.00	2.48	1.45	1.00
10	0.52	0.67	0.82	1.06	1.50	2.29	3.16	2.27	1.42	1.00
20	0.50	0.66	0.80	1.00	1.35	1.92	2.32	1.97	1.38	1.00
30	0.48	0.63	0.77	0.94	1.15	1.48	1.76	1.70	1.32	1.00
40	0.45	0.59	0.72	0.86	1.00	1.21	1.40	1.42	1.22	1.00
50	0.40	0.53	0.66	0.78	0.88	1.05	1.18	1.22	1.15	1.00
60	0.36	0.47	0.59	0.69	0.80	0.92	1.04	1.11	1.09	1.00
70	0.32	0.42	0.52	0.60	0.72	0.84	0.94	1.02	1.05	1.00
80	0.30	0.38	0.46	0.53	0.63	0.75	0.88	0.97	1.00	1.00
90	0.29	0.35	0.42	0.48	0.55	0.65	0.78	0.90	0.97	1.00
100	0.28	0.34	0.40	0.45	0.51	0.58	0.70	0.82	0.94	1.00
110	0.27	0.33	0.38	0.42	0.47	0.52	0.62	0.76	0.90	1.00
120	0.27	0.32	0.36	0.40	0.44	0.49	0.57	0.69	0.85	1.00
130	0.26	0.31	0.35	0.38	0.42	0.46	0.53	0.64	0.82	1.00
140	0.26	0.30	0.34	0.37	0.40	0.44	0.50	0.61	0.79	1.00
150	0.25	0.29	0.33	0.36	0.38	0.42	0.48	0.59	0.76	1.00
160	0.25	0.29	0.32	0.35	0.38	0.41	0.47	0.57	0.75	1.00
170	0.25	0.29	0.32	0.34	0.37	0.41	0.46	0.56	0.74	1.00
180	0.25	0.29	0.32	0.34	0.36	0.40	0.46	0.56	0.74	1.00

AS = Azimuthal separation between the sky patch and sun.

Table 4.5.9 Measured luminance distribution data for intermediate sky, SOLALT=70 degrees (Nakamura et al., 1985)

AS degrees	Altitude angle of the sky patch (degrees)									
	0	10	20	30	40	50	60	70	80	90
0	0.33	0.43	0.54	0.74	0.90	1.07	1.48	2.70	1.54	1.00
10	0.33	0.42	0.53	0.73	0.89	1.05	1.43	1.90	1.48	1.00
20	0.33	0.41	0.51	0.69	0.87	1.02	1.30	1.55	1.35	1.00
30	0.32	0.39	0.48	0.63	0.82	0.97	1.17	1.34	1.26	1.00
40	0.31	0.38	0.45	0.58	0.76	0.93	1.05	1.18	1.19	1.00
50	0.30	0.37	0.43	0.52	0.69	0.86	0.97	1.08	1.10	1.00
60	0.29	0.35	0.40	0.47	0.61	0.79	0.92	1.00	1.05	1.00
70	0.27	0.32	0.38	0.43	0.52	0.70	0.86	0.96	1.00	1.00
80	0.23	0.29	0.35	0.40	0.46	0.57	0.78	0.93	0.99	1.00
90	0.20	0.26	0.32	0.37	0.42	0.48	0.66	0.85	0.96	1.00
100	0.18	0.24	0.29	0.34	0.38	0.44	0.54	0.74	0.91	1.00
110	0.17	0.22	0.27	0.32	0.36	0.41	0.49	0.65	0.87	1.00
120	0.16	0.21	0.26	0.30	0.34	0.39	0.47	0.60	0.82	1.00
130	0.15	0.20	0.25	0.29	0.33	0.37	0.45	0.57	0.78	1.00
140	0.15	0.19	0.24	0.28	0.32	0.36	0.43	0.55	0.75	1.00
150	0.15	0.19	0.23	0.27	0.31	0.35	0.42	0.54	0.73	1.00
160	0.15	0.19	0.23	0.27	0.30	0.34	0.41	0.53	0.72	1.00
170	0.15	0.18	0.23	0.27	0.29	0.34	0.41	0.52	0.72	1.00
180	0.15	0.18	0.22	0.26	0.29	0.34	0.40	0.52	0.72	1.00

AS = Azimuthal separation between the sky patch and sun.

represents only those conditions when the sky is completely covered with thick and dark clouds. However, it is hardly the representative design condition for locations such as those lying within or near the tropical belt. For example, in Japan the frequency of occurrences of clear, intermediate and overcast skies were respectively found to be 5%, 70% and 25% (Nakamura et al. 1985). The CIE standard overcast sky was thus found to be inadequate to describe the real interior daylit environment. It has been argued that, for fuller exploitation of daylight in buildings, prediction tools should be based upon the real rather than a conservative estimate of the sky luminance distribution.

Based on a ten-year measurement programme in Japan, an intermediate sky has been proposed by Nakamura et al. (1985). The varying luminance distributions were found to lie between the CIE standard clear sky and the CIE standard overcast sky. Tables 4.5.3–4.5.9 show the above mentioned measured luminance ratios of the given sky elements and the zenith.

4.5.4 All-sky distributions

Sky light is a non-uniform extended light source. Its intensity and spatial distribution vary as a function of prevailing sky conditions. In addition to direct sunlight, sky luminance angular distribution is the necessary and sufficient information required for calculating daylight penetration into any properly described environment. Because actual sky luminance distribution data are available only in a handful of locations, it is essential to be able to estimate sky light distribution from routine measurements such as irradiance. In this section luminance distributions for all sky conditions are presented.

Perez et al. (1993) have presented an all sky model which is a generalisation of the CIE (1973) standard clear sky formula. This expression includes five coefficients that can be adjusted to account for luminance distributions ranging from totally overcast to clear skies. The relative luminance l_v, defined as the ratio between the sky luminance at the considered point L_v and the luminance of an arbitrary reference point, is given by

$$l_v = f(\theta, \xi) = [1 + a \exp(b/\cos\theta)][1 + c \exp(d\xi) + e \cos\xi^2] \qquad (4.5.6)$$

For $x = a, b, c, d$ and e, using Table 4.5.10, where ε is sky clearness:

for ε ranges 2–8: $x = x_1 + x_2 z + \Delta [x_3 + x_4 z]$ \qquad (4.5.7a)

else,

$$c = \exp\{(\Delta [c_1 + c_2 z])c_3\}^{-1} \qquad (4.5.7b)$$

$$d = -\exp\{\Delta [d_1 + d_2 z]\} + d_3 + \Delta d_4 \qquad (4.5.7c)$$

The coefficients a, b, c, d and e are adjustable functions of irradiance conditions. Table 4.5.10 provides the required information for obtaining the above functions.

Perez et al.'s (1993) experimental data set included more than 16 000 full-sky scans from Berkeley, California covering a wide range of conditions from overcast to clear through intermediate skies. The Perez et al. all-sky model has been incorporated in the RADIANCE software package by the Fraunhofer Institute for Solar Energy Systems in Freiburg, Germany (Sick, 1994) and by Mardaljevic (1996) at DeMontfort University in Leicester, UK. The latter has evaluated the performance of four sky luminance distribution models by comparing their estimates against vertical illuminance measurements. The models evaluated were: CIE overcast sky (Moon and Spencer, 1942); CIE clear sky (CIE, 1973); intermediate sky (Matsuura and Iwata, 1990); and Perez et al. all sky (Perez et al. 1993). The reported results may be enumerated as follows:

(a) The CIE overcast sky model shows an overall negative bias in the predictions of illuminances due to north- and east-facing windows.
(b) The CIE clear sky model is the worst of all four models, with MBEs being three to four times those associated with the intermediate sky model and almost ten times more than the Perez et al. all-sky model.
(c) For all but one aspect the Perez et al. model generates single digit MBEs.

Prog4-3.For and Prog4-4.For, which respectively use the relative and absolute co-ordinate scheme for the sky patch, enable computation of the sky luminance distribution. It must be borne in mind that both of the above mentioned FORTRAN programs require

Table 4.5.10 Coefficients for Perez et al. (1993) all-sky luminance distribution model, Eqs (4.5.6) and (4.5.7)

	Sky clearness ε									
Range	From	To								
			a_1	a_2	a_3	a_4	b_1	b_2	b_3	b_4
1	1.000	1.065	1.3525	−0.2576	−0.2690	−1.4366	−0.7670	0.0007	1.2734	−0.1233
2	1.065	1.230	−1.2219	−0.7730	1.4148	1.1016	−0.2054	0.0367	−3.9128	0.9156
3	1.230	1.500	−1.1000	−0.2515	0.8952	0.0156	0.2782	−0.1812	−4.5000	1.1766
4	1.500	1.950	−0.5484	−0.6654	−0.2672	0.7117	0.7234	−0.6219	−5.6812	2.6297
5	1.950	2.800	−0.6000	−0.3566	−2.5000	2.3250	0.2937	0.0496	−5.6812	1.8415
6	2.800	4.500	−1.0156	−0.3670	1.0078	1.4051	0.2875	−0.5328	−3.8500	3.3750
7	4.500	6.200	−1.0000	0.0211	0.5025	−0.5119	−0.3000	0.1922	0.7023	−1.6317
8	6.200	9999.999	−1.0500	0.0289	0.4260	0.3590	−0.3250	0.1156	0.7781	0.0025
			c_1	c_2	c_3	c_4	d_1	d_2	d_3	d_4
1	1.000	1.065	2.8000	0.6004	1.2375	1.0000	1.8734	0.6297	0.9738	0.2809
2	1.065	1.230	6.9750	0.1774	6.4477	−0.1239	−1.5798	−0.5081	−1.7812	0.1080
3	1.230	1.500	24.7219	−13.0812	−37.7000	34.8438	−5.0000	1.5218	3.9229	−2.6204
4	1.500	1.950	33.3389	−18.3000	−62.2500	52.0781	−3.5000	0.0016	1.1477	0.1062
5	1.950	2.800	21.0000	−4.7656	−21.5906	7.2492	−3.5000	−0.1554	1.4062	0.3988
6	2.000	4.500	14.0000	−0.9999	−7.1406	7.5469	−3.4000	−0.1078	−1.0750	1.5702
7	4.500	6.200	19.0000	−5.0000	1.2438	−1.9094	−4.0000	0.0250	0.3844	0.2656
8	6.200	9999.999	31.0625	−14.5000	−46.1148	55.3750	−7.2312	0.4050	13.3500	0.6234
			e_1	e_2	e_3	e_4				
1	1.000	1.065	0.0356	−0.1246	−0.5718	0.9938				
2	1.065	1.230	0.2624	0.0672	−0.2190	−0.4285				
3	1.230	1.500	−0.0156	0.1597	0.4199	−0.5562				
4	1.500	1.950	0.4659	−0.3296	−0.0876	−0.0329				
5	1.950	2.800	0.0032	0.0766	−0.0656	−0.1294				
6	2.800	4.500	−0.0672	0.4016	0.3017	−0.4844				
7	4.500	6.200	1.0468	−0.3788	−2.4517	1.4656				
8	6.200	9999.999	1.5000	−0.6426	1.8564	0.5636				

the coefficients data file In4-3.Csv which is appended in the CD. This file must be loaded in the user's PC, in the directory where Prog4-3.For and Prog4-4.For are resident. In Prog4-3.For and Prog4-4.For the dimensionless parameter l_v has been normalised against zenith luminance. In the relative scheme, the azimuth of any given sky patch is its angular separation from the sun. The other scheme allows the user to obtain the luminance distribution in the absolute frame. Details of the sky element grid adopted for these computations are shown in Figures 4.5.1 and 4.5.2. This fine resolution grid was adopted in accordance with the measurement scheme followed by Nakamura et al. (1985).

Owing to the large output array, respective electronic data files Lumdist3.Dat and Lumdist4.Dat are generated. These files may then be imported in any spreadsheet packages, e.g. Microsoft Excel or Lotus 1-2-3, and further manipulation may be

HOURLY SLOPE IRRADIATION AND ILLUMINANCE   145

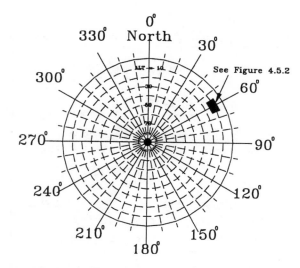

Figure 4.5.1 *Geometry of the sky elements for computation of luminance distribution*

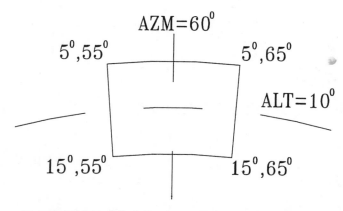

Figure 4.5.2 *Detail of the sky patch shown in Figure 4.5.1*

performed to present the output in the form of a matrix. This procedure is demonstrated via the following example.

Example 4.5.1

In Table 4.5.5 measured luminance distribution data for an intermediate sky for SOLALT = 30 degrees are presented. Use Prog4-3.For to compute the above luminance distribution and compare it against the measured data set.

Table 4.5.11 Comparison of Perez et al. (1993) luminance distribution model against measured data from Japan (Nakamura et al (1985))

AS (degrees)	Nakamura et al. Measured data										Perez et al. Computed data									
	Altitude angle of the sky patch, degrees										Altitude angle of the sky patch, degrees									
	0	10	20	30	40	50	60	70	80	90	0	10	20	30	40	50	60	70	80	90
0	2.09	3.00	5.20	11.50	5.08	2.93	2.04	1.56	1.25	1.00	3.10	4.53	6.61	9.59	6.47	3.81	2.38	1.61	1.18	1.00
10	2.00	2.72	4.20	5.20	4.12	2.86	2.03	1.55	1.24	1.00	2.95	4.20	5.78	6.92	5.55	3.58	2.31	1.59	1.17	1.00
20	1.70	2.24	2.86	3.25	2.95	2.48	1.96	1.53	1.23	1.00	2.60	3.47	4.35	4.73	4.15	3.05	2.13	1.53	1.16	1.00
30	1.52	1.86	2.23	2.56	2.46	2.17	1.86	1.49	1.12	1.00	2.19	2.74	3.19	3.33	3.06	2.49	1.90	1.45	1.14	1.00
40	1.40	1.66	1.89	2.15	2.19	2.02	1.73	1.41	1.18	1.00	1.83	2.16	2.40	2.44	2.30	2.01	1.66	1.35	1.11	1.00
50	1.30	1.53	1.71	1.89	1.94	1.78	1.53	1.29	1.22	1.00	1.56	1.76	1.87	1.88	1.80	1.65	1.45	1.25	1.07	1.00
60	1.19	1.38	1.55	1.66	1.68	1.56	1.34	1.19	1.08	1.00	1.37	1.49	1.54	1.52	1.46	1.38	1.27	1.16	1.04	1.00
70	1.03	1.23	1.39	1.46	1.47	1.38	1.21	1.11	1.05	1.00	1.23	1.31	1.32	1.29	1.24	1.19	1.13	1.07	1.01	1.00
80	0.94	1.09	1.21	1.30	1.28	1.20	1.13	1.07	1.03	1.00	1.14	1.19	1.18	1.14	1.09	1.06	1.03	1.00	0.97	1.00
90	0.87	0.98	1.08	1.15	1.11	1.06	1.04	1.02	1.00	1.00	1.09	1.12	1.09	1.04	0.99	0.96	0.95	0.94	0.94	1.00
100	0.82	0.90	0.95	0.97	0.96	0.95	0.94	0.95	0.97	1.00	1.05	1.07	1.04	0.97	0.92	0.89	0.89	0.89	0.91	1.00
110	0.78	0.82	0.83	0.83	0.84	0.85	0.86	0.90	0.95	1.00	1.03	1.04	1.00	0.93	0.88	0.85	0.84	0.86	0.89	1.00
120	0.75	0.75	0.75	0.74	0.75	0.77	0.80	0.85	0.92	1.00	1.02	1.02	0.98	0.91	0.85	0.81	0.81	0.82	0.87	1.00
130	0.73	0.71	0.70	0.69	0.70	0.72	0.75	0.81	0.89	1.00	1.02	1.01	0.96	0.89	0.83	0.79	0.78	0.80	0.85	1.00
140	0.70	0.69	0.68	0.68	0.68	0.69	0.72	0.79	0.87	1.00	1.02	1.01	0.96	0.88	0.81	0.77	0.76	0.78	0.84	1.00
150	0.68	0.67	0.67	0.67	0.68	0.68	0.71	0.77	0.86	1.00	1.02	1.01	0.96	0.87	0.81	0.76	0.75	0.77	0.83	1.00
160	0.66	0.66	0.66	0.67	0.68	0.67	0.70	0.76	0.85	1.00	1.02	1.01	0.96	0.87	0.80	0.76	0.74	0.76	0.82	1.00
170	0.64	0.65	0.66	0.68	0.68	0.67	0.69	0.76	0.85	1.00	1.02	1.02	0.96	0.87	0.80	0.75	0.74	0.76	0.82	1.00
180	0.63	0.65	0.67	0.68	0.68	0.67	0.69	0.75	0.84	1.00	1.02	1.02	0.96	0.87	0.80	0.75	0.74	0.75	0.82	1.00

AS = Azimuthal separation between the sky patch and sun.

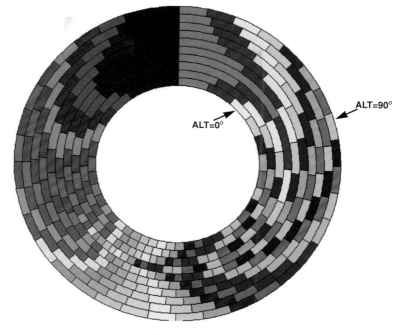

MEASURED

AZIMUTH OF SKY PATCH

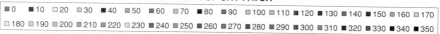

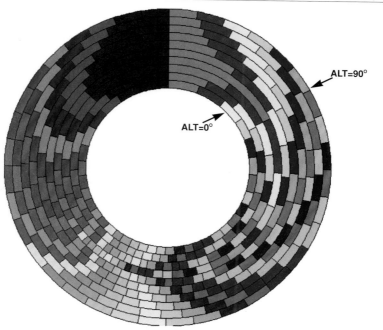

COMPUTED

Figure 4.5.3 *Microsoft Excel doughnut diagram for luminance distribution*

Prog4-3.For is used for this purpose. The routine prompts the user to provide the details of time and location. The luminance distribution l_v is then provided as a function of the sky patch azimuth and altitude. The column-wise output may be imported in Excel and manipulated to produce a matrix table such as Table 4.5.11. The latter table shows good agreement between the measured and computed data sets.

Once the output has been imported in the spreadsheet medium, further exploration may be carried out to visualise the measured and computed sky luminance distribution. This is demonstrated in Figure 4.5.3 which, once again, shows close similarity between the measured and computed results. The doughnut type chart uses arc lengths to demonstrate the variation of the sky luminance, the length of each arc being in direct proportion to the luminosity of the given sky patch.

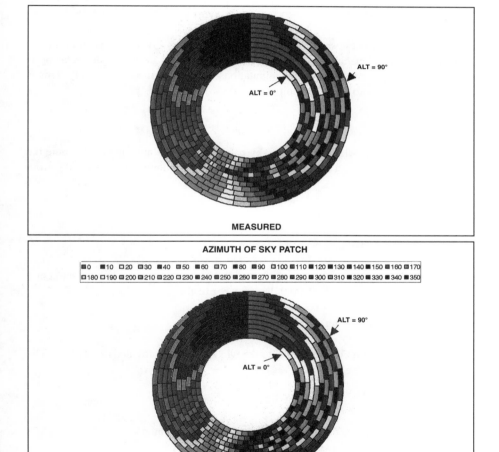

Figure 4.5.3 *Microsoft Excel doughnut diagram for luminance distribution*

4.6 Luminance transmission through glazing

Hopkinson et al. (1966) have presented daylight transmission curves for single, double and triple glazings of clear float glass design. The CIE sky component tables, presented in Chapter 3, incorporate transmission loss for single glazing. For double and triple glazings the BRE (1986) recommends respective correction factors of 0.9 and 0.8 to be multiplied to the total daylight factor. This is certainly simple but is an approximate approach. In the present context, it is possible to investigate the fine detail of daylight penetration through multiple glazings without any loss of serious accuracy by combining the above luminosity estimates with the glazing transmission characteristics. This material is presented in the following sections.

4.6.1 Incidence angle of luminance from a given sky patch

In order to obtain the transmission of luminance from any given sky patch discussed above, the angular separation of the sky patch and a normal to the window glazing is required. Prog4-5.For enables such computations to be performed quite easily. This is demonstrated via the following example.

Example 4.6.1

Refer to Figure 3.6.1, wherein the four sky patches SP1–SP4 are shown. Compute the incidence angle between a normal to the window and the luminance emanating from these sky patches.

Prog4-5.For prompts as follows:

   Enter altitude and azimuth of the first point

At this point the respective altitude and azimuth (measured from true north) of the centre of SP1 is to be provided as 40.0, 90.0. Prog4-5.For further prompts as:

   Enter altitude and azimuth of the second point

The altitude and azimuth of the normal to the window is now provided as 0.0, 95.0. The routine then provides the answer as:

   42.3 (degrees)

Using a similar procedure the incidence angle for SP2 is obtained as 40.3 (degrees). SP3 and SP4 are then obtained from consideration of symmetry.

4.6.2 Transmission of luminance (or radiance) through multiple glazed windows

Prog4-6.For is provided which enables computation of transmission of luminance through multiple glazings. Four types of glazings have been modelled herein, i.e. float glass in single, double and triple panes, and Pilkington Kappafloat double glazing. The latter design, owing to its energy efficiency features, is predominantly being used in new buildings.

Example 4.6.2

Refer to Example 4.6.1 in which the incidence angles of the luminance emanating from the four sky patches, SP1–SP4 were computed. Estimate the luminance transmission through the window shown in Figure 3.6.1. The window may be assumed to be single glazed.

Using the incidence angles obtained in Example 4.6.1, Prog4-6.For enables the required transmissivities to be computed as follows:

Transmissivity τ of luminance emanating from SP1 = 0.905
Transmissivity τ of luminance emanating from SP2 = 0.908

Example 4.6.3

Refer to Table 3.5.4 which provides illuminance data for Watford, North London. Using the Perez et al. all-sky luminance distribution model of Eq. (4.5.6), obtain the internal illuminance for the reference point P shown in Figure 3.6.1 for 0930 hours. Use data obtained in the above two examples.

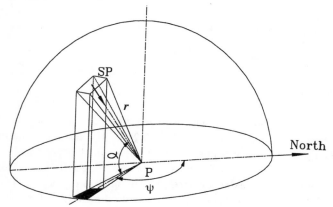

Figure 4.6.1 *Geometry of a given sky patch (SP)*

Figure 4.6.1 shows a sketch of the sky hemisphere and the geometry for any given sky patch SP. A derivation for estimating the illuminance on a horizontal surface due to the given sky patch is now presented. First,

area of sky patch = $(r\, d\alpha)\, (r\, d\psi)$

where $d\alpha$ and $d\psi$ determine the size of the sky patch. In Prog4-3.For and Prog4-4.For these values are each set at 10 degrees ($\pi/18$ radians).

Let $\Delta(IL)$ represent the internal horizontal illuminance at the reference point P due to the given sky patch. The total internal illuminance due to sky may then be obtained by summing the contributions of all relevant sky patches. It may easily be shown that

$$\Delta(IL) = \tau L\ (r\ d\alpha)\ (r\ d\psi) \sin \alpha\ /\ r^2 \qquad (4.6.1)$$

where L is the luminance of the sky patch in cd/m². Using the above fixed values of $\pi/18$ for $d\alpha$ and $d\psi$ in Prog4-3.For and Prog4-4.For,

$$\Delta(IL) = \tau L\ (\pi/18)^2 \sin \alpha \qquad (4.6.2)$$

Based on Examples 4.6.1 and 4.6.2, an average incidence angle of 41.30 and an average τ of 0.906 may be assumed for the four sky patches SP1–SP4. Recall that in the computed Example 3.5.1 (here $L_{z,\ comp}$) was obtained as 5865 cd/m² (see Table 3.5.4). Table 4.5.12 summarises the output from Prog4-3.For and the rest of the computations.

Table 4.5.12 Output for Example 4.6.3

Sky patch	L / L_z	$L_{z,\ comp}$ (cd/m²)	$L_{z,\ comp} \times (L/L_z)$ Prog4-3.For (lx)	$\Delta(IL)$ Eq. (4.6.2)
SP1	1.72	5865	10 090	179
SP2	2.19	5865	12 840	228
SP3	2.90	5865	17 010	302
SP4	3.95	5865	23 170	411
Total internal illuminance				1120

4.7 Exercises

4.7.1 File3-1.Csv contains 5 minute averaged horizontal and slope irradiance as well as illuminance data for Edinburgh (55.95°N, 3.2°W). Use the horizontal irradiance data to compute slope irradiation using the seven all-sky models presented in Section 4.3. Compare your estimates against the measured values using MBE and RMSE statistics as well as scatter plots for the deviations. Comment on the adequacy of the above models using the criteria laid out in Chapter 1. You may use Prog4-1.For for this task.

4.7.2 Using File3-1.Csv perform computations for slope illuminance using the Perez et al. and the Muneer and Angus models. Compare your estimates against the measured values using MBE and RMSE statistics as well as scatter plots for the deviations. Comment on the adequacy of the above models using the criteria laid out in Chapter 1. You may use Prog4-1.For for this task.

4.7.3 Measured luminance distribution data for Japan were given in Tables

4.5.3–4.5.9. Using Prog4-4.For (or Prog4-4.Exe) obtain the computed luminance distributions. Compare the estimates against the reported values.

4.7.4 Refer to Exercise 3.9.8. Compute the transmission of illuminance for all the above mentioned points for which the sky components were obtained using the procedure presented in Example 4.6.3.

References

BRE (1986) *Estimating Daylight in Buildings: Part 1*. BRE Digest 309. Building Research Establishment, Watford.

CIBSE (1982) *CIBSE Guide A2* (1982) Chartered Institution of Building Services Engineers, London.

CIE (1973) *Standardisation of Luminous Distribution on Clear Skies*. International Conference on Illumination, Commission Internationale de l'Éclairage, Paris.

Dogniaux, R. (1954) Étude du climat de la radiation en Belgique. *Inst. R. Met. Belg. Contr.* (18), 54.

Dubayah, R. and Rich, P.M. (1995) Topographic solar radiation models for GIS. *Int. J. Geographic Information Systems* 9, 405.

Fritz, S. (1955) Illuminance and luminance under overcast skies. *J. Opt. Soc. Amer.* 45, 820.

Goudriaan, J. (1977) *Crop Micrometeorology: a Simulation Study*. Centre for Agricultural Publishing and Documentation, Wageningen, The Netherlands.

Grace, J. (1971) The directional distribution of light in natural and controlled environment conditions. *J. Appl. Ecology* 8, 155.

Gueymard, C. (1987) An anisotropic solar irradiance model for tilted surfaces and its comparison with selected engineering algorithms. *Solar Energy* 38, 367. Erratum, *Solar Energy* 40, 175.

Haves, P. (1988) *SERI-RES Building Thermal Simulation Model Version 1.2*. ETSU/HMSO, London.

Haves, P. and Littlefair, P.J. (1988) Daylight in dynamic thermal modelling programs: case study. *BSER&T* 9, 183.

Hay, J.E. (1979) Calculation of monthly mean solar radiation for horizontal and inclined surfaces. *Solar Energy* 23, 301.

Hay, J.E. and Davies, J.A. (1980) Calculation of the solar radiation incident on an inclined surface. *Proc. First Canadian Solar Radiation Data Workshop*, Ministry of Supply and Services, Canada, p. 59.

Hay, J.E. and McKay, D.C. (1988) *Calculation of Solar Irradiances for Inclined Surfaces: Verification of Models which Use Hourly and Daily Data*. IEA Task IX Final Report, Atmospheric Environment Service, Downsview, Canada.

Hopkinson, R.G. (1954) Measurements of sky luminance distribution at Stockholm. *J. Opt. Soc. Amer.* 44, 455.

Hopkinson, R.G., Petherbridge, P. and Longmore, J. (1966) Daylighting. Heinemann, London.

Iqbal, M. (1983) *An Introduction to Solar Radiation*. Academic Press, New York.

Kambezedis, H.D., Psiloglou, B.E. and Gueymard, C. (1994) Measurements and models for total solar irradiance on inclined surface in Athens, Greece. *Solar Energy* 53, 177.

Kimball, H.H. and Hand, I.F. (1921) Sky brightness and daylight illumination measurements. *Mon. Weath. Rev.* 49, 481.

Klucher, T.M. (1979) Evaluation of models to predict insolation on tilted surfaces. *Solar Energy* 23, 111.

Kondratyev, K.Y. (1969) *Radiation in the Atmosphere*. Academic Press, London

Kondratyev, K.Y. and Manolova, M.P. (1960) The radiation balance of slopes. *Solar Energy* 4, 14.

Liu, B.Y.H. and Jordan, R.C. (1960) The inter-relationship and characteristic distribution of direct, diffuse and total solar radiation. *Solar Energy* 4, 1.

Ma, C.C.Y. and Iqbal, M. (1983) Statistical comparison of models for estimating solar radiation on inclined surfaces. *Solar Energy* 31, 313.

Mardaljevic, J. (1995) Validation of a lighting simulation program under real sky conditions. *Lighting Res. & Tech.* 27, 181.

Mardaljevic, J. (1996) Private communication. DeMontfort University, Leicester.

Matsuura, K. and Iwata, T. (1990) A model of daylight source for the daylight illuminance calculations on all weather conditions. *Proc. Third Int. Daylighting Conference*, Moscow.

Meteorological Office (1980) *Solar Radiation Data for the United Kingdom 1951–75*. MO 912. Meteorological Office, Bracknell.

Moon, P. and Spencer, D.E. (1942) Illumination from a non-uniform sky. *Trans. Illum. Eng. Soc. NY* 37, 707.

Muneer, T. (1987) *Solar Radiation Modelling for the United Kingdom*. PhD thesis, Council for National Academic Awards, London.

Muneer, T. (1990a) Further evaluation of the Muneer's solar radiation model. *BSER&T* 11, 77.

Muneer, T. (1990b) Solar radiation model for Europe. *BSER&T* 11, 153.

Muneer, T. (1995) Solar irradiance and illuminance models for Japan I. Sloped surfaces. *Lighting Res. & Tech.* 27, 209.

Muneer, T. and Angus, R.C. (1993) Daylight illuminance models for the United Kingdom. *Lighting Res. & Tech.* 25, 113.

Nagata, T. (1990a) *Radiance Distribution on Overcast Skies*. Internal report, Department of Architecture and Civil Engineering, Fukui University, Fukui, Japan.

Nagata, T. (1990b) Radiance distribution on clear skies. *Trans. Hokuriku Branch of Arch. Inst. of Japan* 33, 213.

Nakamura, H., Oki, M. and Hayashi, Y. (1985) Luminance distribution of intermediate sky. *J. Light & Visual Environment* 9, 6.

Perez, R., Ineichen, P. and Seals R. (1990) Modelling daylight availability and irradiance components from direct and global irradiance. *Solar Energy* 44, 271.

Perez, R., Seals, R. and Michalsky, J. (1993) Modelling skylight angular luminance distribution from routine irradiance measurements. *J. Illum. Eng. Soc.* winter, 10.

Peyre, J. (1927) Mesure de la brillance du ciel diurne. *Revue d'Optique* 6, 73.

Pokrowski, G.I. (1929) Uber einen scheinbaren Mie-Effect und seine mogliche Rolle in der Atmospharenoptik. *A. Phys.* 53, 67.

Rayleigh, Lord J.W.S. (1871) On the light from the sky, its polarisation and colour. *Phil. Mag.* 41, 107.

Reindl, D.T., Beckman, W.A. and Duffie, J.A. (1990) Evaluation of hourly tilted surface radiation models. *Solar Energy* 45, 9.

Saluja, G.S. and Muneer, T. (1987) An anisotropic model for inclined surface solar irradiation. *Proc. Inst. Mech. Engrs.* 201, C1, 11.

Sick, F. (1994) Private communication. Fraunhofer Institute for Solar Energy Systems, Freiburg, Germany.

Skartveit, A. and Olseth, J.A. (1986) Modelling slope irradiance at high latitudes. *Solar Energy* 36, 333.

Steven, M.D. (1977a) Standard distribution of clear sky radiance. *Q. J. Roy. Met. Soc.* 103, 457.

Steven, M.D. (1977b) *Angular Distribution and Interception of Diffuse Solar Radiation.* PhD Thesis, Nottingham University.

Steven, M.D. and Unsworth, M.H. (1979) The diffuse solar irradiance of slopes under cloudless skies. *Q. J. Roy. Met. Soc.* 105, 593.

Steven, M.D. and Unsworth, M.H. (1980) The angular distribution and interception of diffuse solar radiation below overcast skies. *Q. J. Roy. Met. Soc.* 106, 57.

Temps, R.C. and Coulson, K.L. (1977) Solar radiation incident upon slopes of different orientations. *Solar Energy* 19, 179.

Usher, J.R. and Muneer, T. (1989) Case studies in solar radiation modelling. *Math. Comput. Modelling* 12, 1155.

Walsh, J.W.T. (1961) *The Science of Daylight.* Macdonald, London.

Ward, G.J. (1994) The RADIANCE lighting simulation and rendering system. *Computer Graphics Proceedings*, Annual Conference Series.

5 SOLAR SPECTRAL RADIATION

H. Kambezidis

When solar radiation enters the earth's atmosphere it is attenuated by scattering and absorption processes. The scattered radiation is called diffuse radiation. A part of the diffuse radiation is irradiated back to space with the rest reaching the ground. The radiation which arrives at the surface of the earth directly from the sun is called direct or beam radiation. The knowledge of spectral irradiance (direct and diffuse) arriving at the surface of the earth is important for engineering and biological applications. The integration of both diffuse and direct radiation over all wavelengths is called broadband and has been the subject of Section 3.3. In this respect one may refer to the works of Laue (1970), Lorente et al. (1994) and de la Casiniére et al. (1995). The importance of solar spectral estimations has been recognised and several algorithms including those of Kneizys et al. (1980), Bird (1984) and Gueymard (1994) have appeared in the literature.

In this chapter an up-to-date review of the relevant work will be discussed, and based on this a routine for computation of spectral solar irradiance for the cloudless sky will be presented. The routine provides the beam normal, spectral energy calculations to be performed at a resolution of one nanometre (nm). There is a dearth of recent, quality measured spectral data. However, comparison is provided herein against such measurements undertaken at the National Observatory of Athens in Greece.

5.1 Monochromatic solar spectral radiation

If $I_{eon\lambda}$ is the spectral, monochromatic, beam normal solar radiation entering the earth's atmosphere and $I_{Bn\lambda}$ is the energy flux arriving at the earth's surface, then Bouguer's or Lambert's law applies,

$$I_{Bn\lambda} = S\, I_{en\lambda}\, \exp(-k_\lambda\, m) \tag{5.1.1}$$

where k_λ is the monochromatic extinction or attenuation coefficient, m is the relative air mass, and S is the correction factor for the sun–earth distance. An approximate relation for S was given in Chapter 1. Spencer (1971) has, however, provided a more refined model which is given as

$$S = 1.000\,11 + 0.034\,221 \cos \text{DNA} + 0.001\,28 \sin \text{DNA}\\ + 0.000\,719 \cos (2\text{DNA}) + 0.000\,077 \sin (2\text{DNA}) \tag{5.1.2}$$

Here DNA=2π (DN-1) / 365, where DN is the day number during any given year (see Section 1.1.1 for the relevant routine to obtain DN).

An updated distribution $I_{o\lambda}$ has been given by Gueymard (1994) which is included in File5-1.Csv. The sun's rays travelling through the terrestrial atmosphere are subjected to depletion of solar energy by various processes which may be quantified by the attenuation coefficients $k_{j\lambda}$. The most important processes are the absorption of solar energy by ozone, mixed gases (CO, CO_2, O_2, N_2, CH_4 and NO_2) and, water vapour, and scattering by molecules (Rayleigh scattering) or aerosols (Mie scattering). Thus if τ_λ is defined as the monochromatic extinction coefficient,

$$\tau_\lambda = \exp\,[k_{r\lambda}m_r + k_{a\lambda}m_a + k_{o\lambda}m_o + k_{g\lambda}m_g + k_{w\lambda}m_w + k_{n\lambda}m_n)] \qquad (5.1.3)$$

where the indices r, α, o, g, w and n respectively refer to Rayleigh and Mie scattering and to absorption due to ozone, mixed gases, water vapour and NO_2. Siegel and Howell (1981) have shown that Rayleigh scattering occurs when $\pi D/\lambda < 0.6/\eta$, where D is the diameter of the air molecules, λ is the wavelength (μm) and η is the refractive index. However, it has been shown that Mie scattering prevails when $0.6/\eta < \pi D/\lambda < 5$. In the Rayleigh mode, the scattering process is identical in the forward and backward directions, but the scattering is predominantly in the forward direction when Mie scattering occurs.

5.2 Absorption and scattering of solar irradiance

5.2.1 Rayleigh scattering

Rayleigh's theory assumes that all scattering particles are spherical and that they scatter independently of one another, each being less than 0.2λ in diameter. An approximate formula for $k_{r\lambda}$ has been given by Leckner (1978):

$$k_{r\lambda} = 0.008\,735\ \lambda^{-4.08} \qquad (5.2.1)$$

This is applicable for dry air at standard conditions such as the *US Standard Atmosphere* (1976).

For Rayleigh scattering, m_r was given by Gueymard (1994) as

$$m_r = [\cos z + 1.767\,59 \times 10^{-3}\,z\,(94.375\,15 - z)^{-1.215\,63}]^{-1} \qquad (5.2.2)$$

where z is the solar zenith angle in degrees.

The Rayleigh optical thickness $\tau_{r\lambda}$ is thus given as

$$\tau_{r\lambda} = \exp\,(-m_r\,k_{r\lambda}\,p\,/\,p_0) \qquad (5.2.3)$$

where p is the atmospheric pressure (mbar) and p_0 is the reference pressure (1013.25 mbar).

Example 5.2.1

Compute m_r and $\tau_{r0.5}$ for the following conditions: $z = 60°$, $\lambda = 0.5$ μm, $p = 925$ mbar.

Using Eqs (5.2.1)–(5.2.3) the required quantities are obtained:

$m_r = 1.99$, $\tau_{r0.5} = 0.764$

5.2.2 Aerosol (Mie) scattering

Ångström (1929; 1930) proposed a single formula, generally known as Ångström's turbidity formula, for the extinction coefficient of the aerosols:

$$k_{a\lambda} = \beta \lambda^{-\alpha} \tag{5.2.4}$$

where β is called Ångström's turbidity coefficient, usually varying in the range 0–0.5, and α is the wavelength exponent with a typical value of 0.5–2.5. A commonly used value for α is 1.3.

According to Gueymard (1994),

$$m_\alpha = [\cos z + 4.294\,52 \times 10^{-4}\, z(92.248\,49 - z)^{-1.2529}]^{-1} \tag{5.2.5}$$

A simple model for β has been given by McClatchey and Selby (1972):

$$3.912/vis - \beta = 0.55^\alpha (0.011\,62)[0.024\,729\,(vis - 5) + 1.132] \tag{5.2.6}$$

where vis is the visibility in the horizontal direction (km). A standard value for α is 1.3. Table 5.2.1 provides values for vis for given atmospheric conditions.

Table 5.2.1 Turbidity values for various visibilities (after Iqbal, 1983)

Atmospheric condition	α	β	vis (km)
Clean	1.3	0	340
Clear	1.3	0.1	28
Turbid	1.3	0.2	11
Very turbid	1.3	0.4	≤5

Then $\tau_{a\lambda}$ is estimated as

$$\tau_{a\lambda} = \exp[-m_\alpha \beta \lambda^{-\alpha}] \tag{5.2.7}$$

Example 5.2.2

Compute m_α and $\tau_{a0.5}$ for the following given conditions: $\lambda = 0.5$ μm, $vis = 11$ km and $z = 60°$.

158   SOLAR RADIATION AND DAYLIGHT MODELS

Using Eqs (5.2.5) and (5.2.7), the required quantities are obtained as

$m_\alpha = 2, \quad \beta = 0.2, \quad \tau_{\alpha 0.5} = 0.37$

5.2.3 Absorption by ozone

The spectral transmittance for ozone $\tau_{o\lambda}$ is expressed in the form

$$\tau_{o\lambda} = \exp[-(k_{o\lambda}\, m_o\, \{l_o/1000\})] \tag{5.2.8}$$

where the total ozone column in the atmosphere l_o (milli-atm-cm) and m_o are respectively obtained using the formulation (Eq.1 3.3.1) due to Van Heuklon (1979) and the following due to Gueymard (1994):

$$m_o = [\cos z + 1.074\,89 \times 10^{-2} z (96.626\,67 - z)^{-1.3882}]^{-1} \tag{5.2.9}$$

Ozone absorbs strongly in the ultraviolet waveband, moderately in the visible band and only weakly in the near infrared. Depending on temperature T (K), $k_{o\lambda}$ is obtained via the Smith et al. (1992) formulation

$$k_{o\lambda}(T_{eo}) = \max\,[0, k_{o\lambda}(T_{ro}) + C_1(T_{eo} - T_{ro}) + C_2(T_{eo} - T_{ro})^2] \tag{5.2.10a}$$

where $T_{eo} = a_0 + a_1 T$ ($a_0 = 332.41$ K, $a_1 = -0.344\,67$ for summer; $a_o = 142.68$ K, $a_1 = 0.284\,98$ for winter) characterises any given temperature which may be other than the reference temperature T_{ro} for which the ozone absorption coefficient must be calculated. The C coefficients are obtained as follows:

$C_1 = (0.253\,26 - 1.7253\,\lambda + 2.9285\,\lambda^2)/(1 - 3.589\,\lambda)$, $\qquad \lambda < 310$ nm
$C_2 = (9.6635 \times 10^{-3} - 6.3685 \times 10^{-2}\,\lambda + 0.104\,64\,\lambda^2)/(1 - 3.6879\,\lambda)$ $\qquad \lambda < 310$ nm
$C_1 = (0.396\,26 - 2.3272\,\lambda + 3.4176\,\lambda^2)/(1 - 3.5\,\lambda)$, $\qquad 310 \le \lambda \le 344$ nm
$C_2 = (1.8268 \times 10^{-2} - 0.109\,28\,\lambda + 0.163\,38\,\lambda^2)/(1 - 3.5\,\lambda)$, $\qquad 310 \le \lambda \le 344$ nm

The reference temperature $T_{ro} = 228$ K, and $k_{o\lambda}$ is a function of T_{ro}. In the wavelength range $344 < \lambda \le 560$ nm,

$$k_{o\lambda}(T_{eo}) = \max\,\{0, k_{o\lambda}(T_{ro})[1 + 0.003\,708\,3(T_{eo} - T_{ro})e^{28.04(0.4474 - \lambda)}]\} \tag{5.2.10b}$$

For $\lambda > 560$ nm, $k_{o\lambda}$ is read from File5-1.Csv (column 5).

Example 5.2.3

Compute m_o and $\tau_{o0.5}$ for the following conditions: $z = 60°$, $\lambda = 0.5$ μm, $T = 300$ K, geographical latitude = 37.97°N, geographical longitude = 23.72°E, day number DN = 100.

SOLAR SPECTRAL RADIATION   159

Using Eqs (5.2.9) and (5.2.10) the required quantities are obtained as

$m_o = 1.98$, $\tau_{o0.5} = 0.983$

5.2.4 Absorption by mixed gases

The uniformly mixed gases (principally O_2 and CO_2) decrease monotonically in their atmospheric concentration with altitude. According to Pierluissi and Tsai (1986; 1987) the mixed gas transmittance is defined as

$$\tau_{g\lambda} = \exp[-(k_{g\lambda}\, l_g\, m_g)^C] \qquad (5.2.11)$$

According to Gueymard (1994),

$$m_g = [\cos z + 1.767\,59 \times 10^{-3} z(94.375\,15 - z)^{-1.215\,63}]^{-1} \qquad (5.2.12)$$

where

$l_g = C_0\,(p/p_0)^{C_1}\,\theta^{C_2}$, $\theta = 288.15/T$

$C_0 = 4.9293$ km, $C_1 = 1.8849$, $C_2 = 0.1815$ for O_2

$C_0 = 4.8649$ km, $C_1 = 1.9908$, $C_2 = -0.697$ for CO_2

Below 1 µm wavelength l_g is estimated on the assumption that the atmosphere consists of pure O_2. For longer wavelengths l_g is computed under the assumption that CO_2 fills the atmospheric space. $C = 0.5641$ for $\lambda < 1$ µm and $C = 0.707$ for $\lambda \geq 1$ µm. The values of $k_{g\lambda}$ are taken from File5-1.Csv (column 4).

Example 5.2.4

Compute m_g and $\tau_{g0.5}$ for the following conditions: $z = 60°$, $\lambda = 0.5$ µm, $p = 1000$ mbar, $T = 280$ K.

Using Eqs (5.2.11) and (5.2.12) the required quantities are obtained as

$m_g = 1.99$, $\tau_{g0.5} = 1$

5.2.5 Absorption by water vapour

Pierluissi et al. (1989) have proposed the following expression for the water vapour transmittance:

$$\tau_{w\lambda} = \exp-[k_{w\lambda}\,(m_w l_w)^{1.05}\,f_w^n\,B_w]^C \qquad (5.2.13)$$

where m_w may be obtained via Gueymard's (1994) formulation

$$m_w = [\cos z + 4.294\,52 \times 10^{-4}\,z(92.248\,49 - z)^{-1.2529}]^{-1} \quad (5.2.14)$$

The total precipitable water l_w (cm) is obtained using the technique described in Section 3.3.2.3.

In Eq. (5.2.13), C and n are wavelength dependent exponents and B_w is a correction factor for the absorption process. These may be obtained as follows:

$C = 0.538\,51 + 0.003\,262\lambda + 1.5244\,e^{-4.2892\,\lambda}$
$n = 0.886\,31 + 0.025\,274\lambda - 3.5949\,e^{-4.5445\lambda}$
$B_w = f\exp(0.1916 - 0.0785m_w + 4.706 \times 10^{-4}m_w^2)$
$f = 0.624 m_w l_w^{0.457}$ for $k_{w\lambda} < 0.01$
$f = (0.525 + 0.246 m_w l_w)^{0.45}$ for $k_{w\lambda} \geq 0.01$

$K_{w\lambda}$ is the spectral attenuation coefficient for water vapour and its values are available in File 5–1.Csv. The remaining terms of Eq. (5.2.13) are as follows:

$f_w = A_w\,[0.394 - 0.269\,46\lambda + (0.464\,78 + 0.237\,57\lambda)p/p_0]$
where $A_w = 1$ for $\lambda \leq 0.67\,\mu m$, else
$A_w = (0.984\,49 + 0.023\,882\,\lambda)\,l_w^q$, $q = -0.024\,54 + 0.037\,533\lambda$.

Example 5.2.5

Compute m_w and $\tau_{w0.5}$ for the following conditions: $z = 60°$, $p = 1000$ mbar, $\lambda = 0.5\mu m$, $T = 280$ K and RH = 50%.

Using Eqs (5.2.13) and (5.2.14) the required quantities are obtained as

$m_w = 2$, $\tau_{w\,0.5} = 1$

5.2.6 Absorption by nitrogen dioxide

Like ozone, NO_2 transmittance is given by

$$\tau_{n\lambda} = \exp[-(k_{n\lambda}\,l_n\,m_n)] \quad (5.2.15)$$

where (Gueymard, 1994)

$$m_n = [\cos z + 8.47 \times 10^{-3}\,z(96.626\,67 - z)^{-1.388\,02}]^{-1} \quad (5.2.16)$$

and l_n is the total column of NO_2 in the atmosphere (atm-cm). A typical value of l_n is 1.66×10^{-3} atm-cm (Schröder and Davies, 1987). Finally,

$$k_{n\lambda}(T_{en}) = \max\{0, k_{n\lambda}(T_{rn})[1 + (T_{en} - T_{rn})\sum_{i=0}^{5} f_i \lambda^i]\}$$

where $T_{\rm rn} = 243.2$ K (reference temperature) and $T_{\rm eo}$ is defined and estimated as for ozone in Section 5.2.3. Here

for $\lambda < 0.625$ μm:
$f_0 = 0.697\ 73$, $f_1 = -8.182\ 90$, $f_2 = 37.821$, $f_3 = -86.136$, $f_4 = 96.615$, $f_5 = -42.635$
else:
$f_0 = 0.035\ 39$, $f_1 = -0.049\ 85$, $f_2 = f_3 = f_4 = f_5 = 0$

$k_{n\lambda}$ is read from File5-1.Csv (column 6).

NO_2 is a highly variable atmospheric constituent that plays an important role in the ozone cycle. NO_2 and O_3 are natural constituents in the stratosphere, while in the troposphere they imply air pollution. High concentrations of NO_2 in large cities are responsible for the typical brown colour of the pollution cloud (Husar and White, 1976).

Example 5.2.6

Compute m_n and $\tau_{n0.5}$ for the following conditions: $z = 60°$, $\lambda = 0.5$μm, $T = 280$ K and $l_n = 1.66 \times 10^{-3}$ atm-cm.

Using Eqs (5.2.15) and (5.2.16) the required quantities are obtained as

$m_n = 1.99$, $\tau_{n0.5} = 0.985$.

5.3 Spectral radiation model

The spectral radiation model (SRM) is the culmination of the work of several authors described in Sections 5.1 and 5.2. The SRM calculates the beam normal, spectral irradiation under cloudless skies in the wavelength band 300–1700 nm. Most of the energy receipt is contained in this bandwidth. The calculations are performed at a resolution of 1 nm. Some of the air mass and transmission models presented in the above sections have one limitation, i.e. they return unreasonably high values at low solar altitudes. The reader's attention is therefore drawn to the solution incorporated in this text which is to exclude from computations those instances when the solar altitude is less than 7 degrees.

The basic equation is a combination of Eqs (5.1.2) and (5.1.3):

$$I_{Bn\lambda} = S\, I_{En\lambda}\tau_\lambda = S\, I_{En\lambda}\, \tau_{r\lambda}\, \tau_{a\lambda}\, \tau_{o\lambda}\, \tau_{g\lambda}\, \tau_{w\lambda}\, \tau_{n\lambda} \qquad (5.3.1)$$

for a single λ, or

$$I_{Bn} = \sum_{\lambda=\lambda_1}^{\lambda_2} I_{Bn\lambda} = S\sum_{\lambda=\lambda_1}^{\lambda_2} I_{Eon\lambda}\, \tau_{r\lambda}\, \tau_{a\lambda}\, \tau_{o\lambda}\, \tau_{g\lambda}\, \tau_{w\lambda}\, \tau_{n\lambda} \qquad (5.3.2)$$

for the entire region of 300 nm $\leq \lambda_1 \leq \lambda \leq \lambda_2 \leq 1700$ nm.

A routine for the SRM is provided in Prog5-1.For (and Prog5-1.Exe) which enables computation of spectral solar irradiance under cloudless skies. To enable execution the following input data are required in addition to File5-1.Csv:

(a) geographical latitude and longitude
(b) day number for the year
(c) atmospheric pressure (mbar)
(d) ambient temperature (°C)
(e) relative humidity (%)
(f) solar altitude (degrees)
(g) reported or estimated visibility (km).

Example 5.3.1

On 24 May 1995, 1120 h LCT, solar spectral measurements were undertaken at the National Observatory of Athens in Greece under a cloudless sky. The spectral radiometer measurement window was 300–1000 nm. The input data set for the SRM is:

(a) LAT = + 37.97° (north), LONG = − 23.72° (east)
(b) DN = 144
(c) p = 1001.3 mbar
(d) T = 25.4 °C
(e) RH = 42%
(f) SOLALT = 68.2 degrees
(g) observed visibility = 20 km.

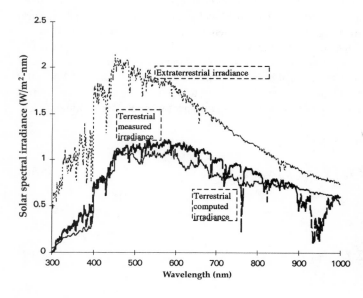

Figure 5.3.1 *Comparison of computed and measured spectral irradiance at Athens, Greece*

The SRM, given in Prog5-1.For computes $I_{Bn\lambda}$ in the specified wavelength region. Figure 5.3.1 shows the extraterrestrial spectrum (for reference), the spectral measurements and the computed SRM curve.

References

Ångström, A. (1929) On the atmospheric transmission of sun radiation and on dust in the air. *Geografis. Annal.* 2, 155.

Ångström, A. (1930) On the atmospheric transmission of sun radiation. *Geograf. Annal.* 2, 130.

ASHRAE (1993) *Handbook of Fundamentals*. American Society of Heating, Refrigerating and Air-Conditioning Engineers, Atlanta.

Bird, R.E. (1984) A simple, solar spectral model for direct-normal and diffuse horizontal irradiance. *Solar Energy* 32, 461.

De la Casiniére, A., Cabot, T. and Benmansour, S. (1995) Measuring spectral diffuse solar irradiance with non-cosine flat-paper diffusers. *Solar Energy* 54, 173.

Gueymard, C. (1994) *Simple Model of the Atmospheric Radiative Transfer of Sunshine (SMARTS2): Algorithms and Performance Assessment*. FSEC-PF-270-94, Florida Solar Energy Center, Cocoa, FL.

Husar, R.B. and White, W.H. (1976) On the colour of the Los Angeles smog. *Atmos. Environ.* 10, 199.

Iqbal, M. (1983) *An Introduction to Solar Radiation*. Academic Press, Toronto.

Kneizys, F.X., Shette, E.P., Gallery, W.O., Chetwynd, J.H., Abren, L.W., Selby, J.E.A., Fenn, R.W. and McClatchey, R.A. (1980) *Atmospheric Transmittance/Radiance: Computer Code LOWTRAN5*. Report. AFGL TR-80-0067, Air Force Geophysics Lab., Hanscom AFB, MA.

Laue, E.G. (1970) The measurement of solar spectral irradiance at different terrestrial elevations. *Solar Energy* 13, 43.

Leckner, B. (1978) The spectral distribution of solar radiation at the earth's surface: elements of a model. *Solar Energy* 20, 143.

Lorente, J., Redano, A. and de Cabo, X. (1994) Influence of urban aerosol on spectral solar irradiance. *J. Appl. Meteorol.* 33, 406.

McClatchey, R.A. and Selby, J.E. (1972) *Atmospheric Transmittance from 0.25 to 38.5 µm: Computer Code LOWTRAN2*. AFCRL-72-0745, Environ. Res. Paper 427.

Pierluissi, J.H. and Tsai, C.M. (1986) Molecular transmittance band model for oxygen in the visible. *Appl. Opt.* 25, 2458.

Pierluissi, J.H. and Tsai, C.M. (1987) New LOWTRAN models for the uniformly mixed gases. *Appl. Opt.* 26, 616.

Pierluissi, J.H., Maragoudakis, C.E. and Tehrani-Morahed, R. (1989) New LOWTRAN band model for water vapour. *Appl. Opt.* 28, 3792.

Schröder, R. and Davies, J.A. (1987) Significance of nitrogen dioxide in estimating aerosol optical depth and the distributions. *Atmosphere-Ocean* 25, 107.

Siegel, R. and Howell, J.R. (1981) *Thermal Radiation Heat Transfer*. McGraw-Hill, New York.

Smith, E.U.P., Wan, Z. and Baker, K.S. (1992) Ozone depletion in Antarctica: modeling its effect on solar UV irradiance under clear-sky conditions. *J. Geoph. Res* 97C, 7383.

Spencer, J.W. (1971) Fourier series representation of the position of the sun. *Search* 2, 172.

US Standard Atmosphere (1976) US Government Printing Office, Washington, DC.

Van Heuklon, T.K. (1979) Estimating atmospheric ozone for solar radiation models. *Solar Energy* 22, 63.

6 GROUND ALBEDO

Solar radiation incident on vertical and inclined surfaces consists of beam, sky-diffuse and ground-reflected components. The ground-reflected component may be significant particularly in the northern latitudes owing to low elevations of the sun and, at times, the presence of highly reflecting snow cover. Accurate estimation of ground-reflected radiation would require knowledge of the foreground type and geometry, its reflectivity, its degree of isotropy, the details of the surrounding skyline and the condition of the sky. Little information is available on the interaction of these parameters. The usual approach is to take a constant value of ground reflectance of 0.2 for temperate and humid tropical localities and 0.5 for dry tropical localities (CIBSE, 1982). This is despite the fact that the reflectivity of snow-covered ground could be as high as 0.9.

The term 'ground albedo' or simply 'albedo' is often used interchangeably with 'ground reflectance'. On the other hand, as Monteith (1959) has pointed out, the term 'albedo' or 'whiteness' refers to the reflection coefficient in the visible range of the spectrum, whereas 'reflectance' denotes the reflected fraction of short-wave energy. In this book the term 'albedo' has been used synonymously with reflectance, applying to the total short-wave energy.

The importance of knowing the albedo for the determination of radiation balance of macro- and microclimates is well known. A good estimate of albedo of the surrounding terrain is a prerequisite for representative calculations related to the energy balance of vegetation, amount of potential transpiration, energy interception of walls, windows, roofs and solar energy collectors. Therefore the small and large scale variation of albedo is of interest. The variation in albedo is spatial and temporal owing to the changing landscapes of the earth and to the seasonal presence of snow and to some extent moisture deposition.

There has been some initiative in the past to assess the spatial variability of albedo. Barry and Chambers (1966) have presented a map of summer albedo over England and Wales. The map was prepared on the basis of published values of albedo and data collected at specific sites by means of ground and airborne measurements over southern England.

The aim of this chapter is to present data on albedo values for a variety of surfaces which may be in the view of solar energy collection systems and to present techniques for computation of averaged albedo values for any locality. In this respect preliminary maps for the UK are also provided which would be useful in estimating the albedo of any land mass for the summer and winter months.

6.1 Estimation of ground-reflected radiation

The two main problems in estimating the ground-reflected radiation are the uncertainty of average reflectance of the neighbouring ground and the lack of an accurate model. Determination of a reasonably accurate value of albedo is not easy. Furthermore, it is more difficult to specify reflectivity than the other radiative properties such as absorptivity. As indicated by Siegel and Howell (1972), no less than eight types of reflectivity are in current use: bidirectional spectral; directional-hemispherical spectral; hemispherical-directional spectral; hemispherical spectral; bidirectional total; directional-hemispherical total; hemispherical-directional total; and hemispherical total. Of these, hemispherical total reflectivities are sufficient for applications such as meteorology and solar heating design, while spectral reflectivities are required for other applications such as electricity generation by photovoltaic cells, photobiology and solar controlled glazing. Several references, including Greiger (1966), Kondratyev (1969), Hay (1970), Eagleson (1970), Meteorological Office (1972), Monteith (1973), Iqbal (1983), Gueymard (1987) and Saluja and Muneer (1988), quote spectral and total values of reflectivities for different landscapes.

Table 6.1.1 Albedo of soil covers

Soil	Albedo (%)
Black earth, dry	14
Black earth, moist	8
Grey earth, dry	25–30
Grey earth, moist	10–12
Ploughed field, moist	14
White sand	34–40
River sand	43
Light clay earth (levelled)	30–31

Table 6.1.3 Albedo of natural surfaces

Surface	Albedo (%)
Fresh snow cover	75–95
Old snow cover	40–70
Rock	12–15
Densely built-up areas	15–25
High dense grass	18–20
Sea ice	36–50
Water surfaces, sea	3–10
Lawn: high sun, clear sky	23
Lawn: high sun, partly cloudy	23
Lawn: low sun, clear sky	25
Lawn: overcast day	23
Dead leaves	30

Table 6.1.2 Albedo of vegetative covers

Class of vegetation	Species at maximum ground cover	Albedo (%)
Farm crops	Grass	24
	Wheat	26
	Tomato	23
	Pasture	25
Natural vegetation and forests	Heather	14
	Bracken	24
	Deciduous woodland	18
	Coniferous woodland	16

Table 6.1.4 Albedo of building materials

Surface	Albedo (%)
Weathered concrete	22
Weathered blacktop	10
Bituminous and gravel roof	13
Crushed rock surface	20
Building surfaces, dark (red brick, dark paints, etc.)	27
Building surfaces, light (light brick, light paints, etc.)	60

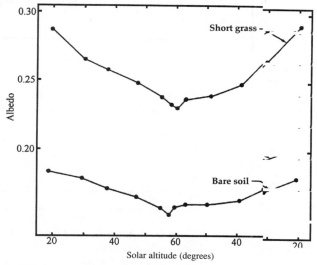

Figure 6.1.1 *Variation of albedo of bars soil and short grass with solar altitude*

Tables 6.1.1–6.1.4 present average hemispherical total albedo for a number of ground coverings. Kondratyev (1969) has highlighted the possible sources of error in the measurement of albedo. Weather and time dependent variations of albedo are illustrated in Figures 6.1.1–6.1.4. Figure 6.1.1 shows the variation of albedo of short grass and bare soil as a function of solar altitude and Figure 6.1.2 shows the albedo of water as a function of cloudiness and solar altitude. Figure 6.1.3 shows the effect of ageing in the albedo of snow and Figure 6.1.4 the variation of that of a snow-covered surface as a function of accumulated temperature index, which is the cumulative value of maximum daily temperatures since the last snowfall.

Hay (1970) has prepared maps of mean monthly albedo for Canada by using the values of albedo reported above and knowledge of a specific dominant surface cover. On the basis of these maps Hay (1976) has tabulated average monthly values of albedo for a number of Canadian locations. Iqbal (1983) has extended Hay's tables to further North American locations.

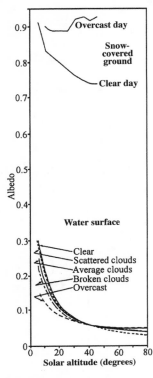

Figure 6.1.2 *Variation of albedo of water surface and snow-covered ground with solar altitude and cloud cover*

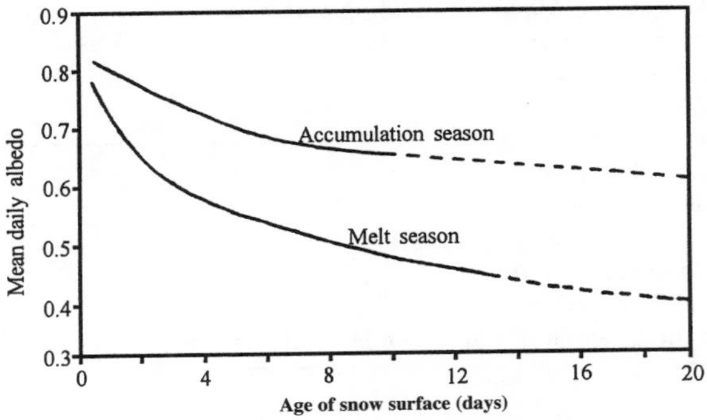

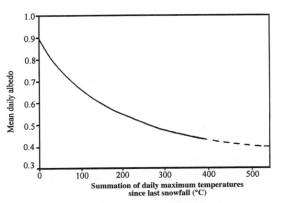

Figure 6.1.3 *Effect of ageing of snow on albedo*

Figure 6.1.4 *Variation of albedo of a snow surface with accumulated temperature index since the last snowfall*

Kung et al. (1964) carried out a series of 12-monthly flights along a selected path in Wisconsin and a series of four long-range flights over extensive areas of North America. From these measurements they produced three continental albedo maps for North America covering winter, summer and transitional seasons. Barry and Chambers (1966) have produced a map of summer albedo for England and Wales.

Tables 6.1.1–6.1.4 characterise the measured values of albedo of small underlying surfaces. However, in engineering applications, such as solar heating design, a weighted average albedo of a relatively large area of foreground is required. Complication arises as this foreground may consists of patches of different ground cover, each with its own characteristic albedo. In their study of the long-term performance of flat-plate solar collectors, Liu and Jordan (1963) adopted a simple approach by assuming an albedo of 0.2 when the ground was covered with less than 25 mm of snow and 0.7 for snow coverage of 25 mm. They calculated, for each month, an average albedo by using a weighting factor equal to the fractional time during the month when the snow depth exceeded 25 mm. Hunn and Calafell (1977) have questioned the adequacy of such a broad assumption and have used photographic methods to compute the weighted average albedo of a few winter landscapes for some urban and rural areas of the USA. They concluded that an average albedo of 0.6–0.7, similar to that used by Liu and Jordan, is applicable to rural landscapes in winter where snow cover is prominent, except for locations adjacent to water surfaces where a considerably lower value of albedo is obtained. For urban areas, Hunn and Calafell maintain that no characteristic albedo in winter may be specified owing to the possibility of wide variations in landscape details, and its value may vary in the range 0.16–0.49. The field of view in snow cover strongly influences the ground reflectivity.

6.2 Models for ground-reflected radiation

Liu and Jordan (1963) have proposed a simple isotropic model for diffusely reflected radiation from the ground. For a surface inclined at an angle TLT to the horizontal and for infinite horizontal foreground, the geometric view factor for radiation exchange between the foreground, and the surface is $\sin^2$ (TLT/2). Without any shadow in the foreground, Liu and Jordan's model results in hourly $I_{g,TLT}$ and daily g_{TLT} reflected radiation on a sloping surface given by

$$I_{g,TLT} = \rho\, I_G \sin^2 (TLT/2) \qquad (6.2.1)$$
$$g_{TLT} = \rho\, G \sin^2 (TLT/2) \qquad (6.2.2)$$

where I_G and G are hourly and daily global irradiation on a horizontal surface and ρ is the average albedo of the foreground.

This model provides an easy means of computing the ground-reflected component; however, it does not take into account the anisotropic nature of the reflected radiation. Furthermore, as illustrated by Figures 6.1.1 and 6.1.2, the hemispherical total reflectivity is a function of the solar elevation. Therefore an adequate model ought to incorporate the following effects:

(a) the variation of albedo of the surfaces in the foreground as a function of solar elevation and cloudiness
(b) the geometric view factors subtended by the surfaces in the foreground on the inclined plane
(c) the details of surrounding skyline in order to estimate the shaded fraction of the foreground.

In addition, as Kondratyev (1969) has shown, all surfaces are to some degree selective as to the wavelength of the incident radiation. Thus it is obvious that precise estimation of reflected radiation is a complex task. Temps and Coulson (1977) have reported the angular dependence of grass-reflected radiation. Based on their measurements for one location in California they have proposed the following model for grass-reflected radiation:

$$I_{g,\text{TLT}} = \rho\, I_G \sin^2(\text{TLT}/2)\,\{[1+\sin^2(z/2)]\cos^2\text{SOLALT}\} \tag{6.2.3}$$

where z is the zenith angle and SOLALT is the solar altitude. This model has only been tested against short-term data on cloudless days.

Gueymard (1987) has proposed more complex models by considering specifically two cases according to the optical characteristics of the reflecting surfaces: the near isotropic reflectance and the pure specular reflectance. The former is of general interest. In this case the reflected radiation can be treated as isotropic in the two half-hemispheres involved. Gueymard has proposed the use of the apparent reflectance ρ_a as given by

$$\rho_a = f_B\, \rho_B\, K_B + \rho_D(1 - K_B) \tag{6.2.4}$$

where f_B is the shadow factor for the beam irradiance, ρ_B is the beam reflectance, K_B the hourly horizontal beam fraction and ρ_D is the reflectance of the diffuse irradiance. ρ_D is a function of the reflectance of the normal incidence beam irradiance ρ_N. ρ_D is given by

$$\rho_D = \rho_N + 0.023(f_{ab} + f_{af}) \tag{6.2.5}$$

where f_{ab} and f_{af} are the anisotropy coefficients which describe the backward and forward reflectance.

Gueymard (1987) has suggested tentative values of the anisotropy coefficients for grass, snow, concrete and other common surfaces.

6.3 Albedo atlas for the United Kingdom

An accurate estimate of albedo for the surrounding terrain is desirable for representative calculations related to the heat balance of vegetation, the potential amount of transpiration, and the energy intercepted by windows and other building surfaces and solar energy collectors. The aim of the present section is to provide preliminary UK maps which would be useful in estimating the albedo of any land mass during the summer and winter months. The methodology presented here may be used with equal effectiveness for other geographical regions.

Snow-covered ground significantly affects the albedo, and hence the incidence of ground-reflected radiation on inclined and vertical surfaces. During snow-free periods albedo is a function of the dominant ground cover and the nature of the surfaces surrounding a building element or a solar collection device. By using the maps of snow cover and monthly average albedo for snow-free periods, a significant improvement can be made in estimating ground-reflected radiation on inclined and vertical surfaces.

Any detailed study of solar radiation on inclined surfaces would require hourly or daily albedo values. Long-term predictions are usually made by employing monthly averages of daily or hourly irradiation values. As a first step, this study aims at the estimation of average monthly albedo for the UK locations. Albedo related maps for winter months have been presented. These maps are reproduced from the work of Muneer (1987).

A number of Meteorological Office stations across the UK collect snow data. Published data are available from detailed measurements covering a limited number of locations. The extent of snow cover is observed every day at 0900 GMT. The depth of snow is also measured by ruler in an area of level snow. For most countries in the temperate belt periodically published climatological memoranda give statistical data of snow depth and days with snow lying at various locations. In the UK such information is available from the Meteorological Office. With reference to the latter source, 'a day of snow falling' is designated if snow or sleet is observed to fall at the station at any time during the 24 hour period, while 'a day of snow lying' is defined as one in which snow covers at least half the ground surrounding the observing station at 0900 GMT.

Kung et al. (1964) have shown that the albedo of snow-covered ground is related to the depth of snow. The snow depth data for the UK are only readily available on an annual basis. In the absence of monthly snow depth data, UK maps are reproduced for the mean number of days on which the snow was lying on the ground for the months of December to March.

The UK maps for November and April were prepared by Muneer (1987) using the following procedure. A correlation was established between snow falling and snow lying on the ground for the months of December and March. Using the maps of snow falling in November and April and employing the correlation for December and March, respectively, maps of snow lying for November and April were prepared.

The contours of equal numbers of days when the snow was lying on the ground throughout the UK are shown in Figures 6.3.1–6.3.6 for the months between November and April. The rest of the period is considered to be free of snow.

6.4 Estimation of monthly-averaged albedo

Following the procedure suggested by Liu and Jordan (1963) the monthly-averaged albedo $\bar{\rho}$ for any month is given by

$$\bar{\rho} = \rho_{\text{snow-free}} (1 - f_{\text{snow}}) + \rho_{\text{snow}} f_{\text{snow}} \qquad (6.4.1)$$

where f_{snow} is the fractional time during the month when the snow was lying on the ground.

Liu and Jordan (1963) used the period when the snow depth was over 25 mm. A snow albedo of 0.7 (see Figure 6.1.3) may be used for snow which is two to three days old. The albedo of grass is in the range 0.23–0.25 (see Tables 6.1.2 and 6.1.3). Monteith (1973), using measurements in the UK, has adopted 0.24 for the albedo value for grass. The albedo for heather has been quoted as between 0.1 and 0.14 while the range of albedo for urban areas has been taken as 0.15–0.25 (see Table 6.1.3). Kung et al. (1964) found the measured albedo for the snow-free period as 0.13–0.17 for the suburban area of Madison, Wisconsin and 0.15–0.18 in the downtown area. Their findings are in agreement with those of Oguntoyinbo (1970) for Nigeria. Saluja and Muneer (1988) have suggested the corresponding value for the UK as 0.14–0.18.

172 SOLAR RADIATION AND DAYLIGHT MODELS

Figure 6.3.1 *Mean number of days with snow lying at 0900 GMT for November*

GROUND ALBEDO 173

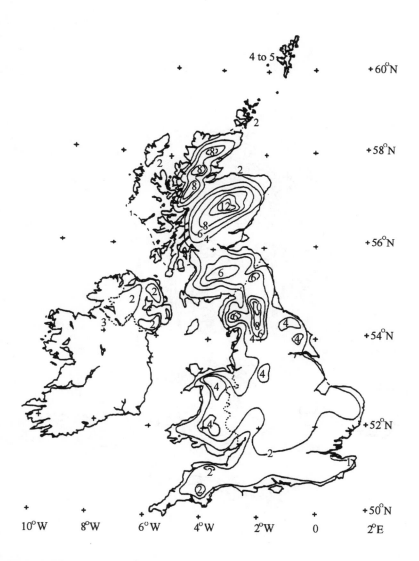

Figure 6.3.2 *Mean number of days with snow lying at 0900 GMT for December*

174 SOLAR RADIATION AND DAYLIGHT MODELS

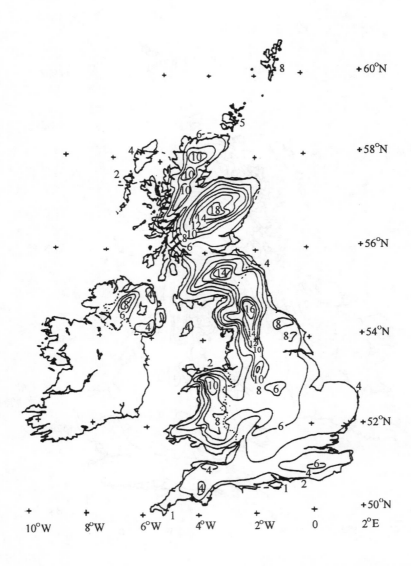

Figure 6.3.3 *Mean number of days with snow lying at 0900 GMT for January*

GROUND ALBEDO 175

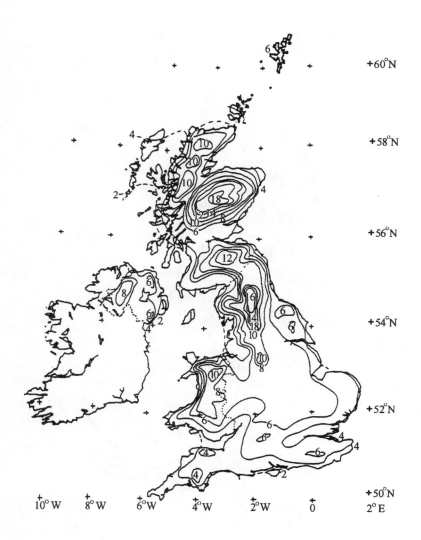

Figure 6.3.4 *Mean number of days with snow lying at 0900 GMT for February*

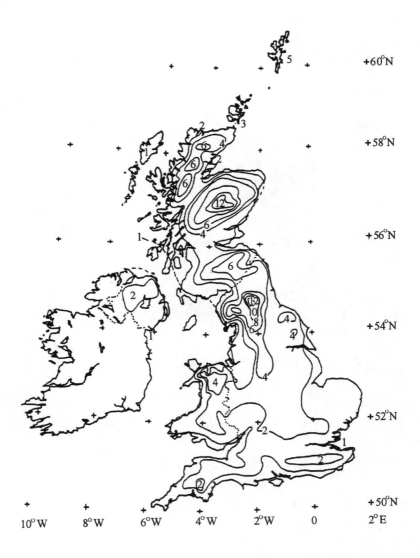

Figure 6.3.5 *Mean number of days with snow lying at 0900 GMT for March*

GROUND ALBEDO 177

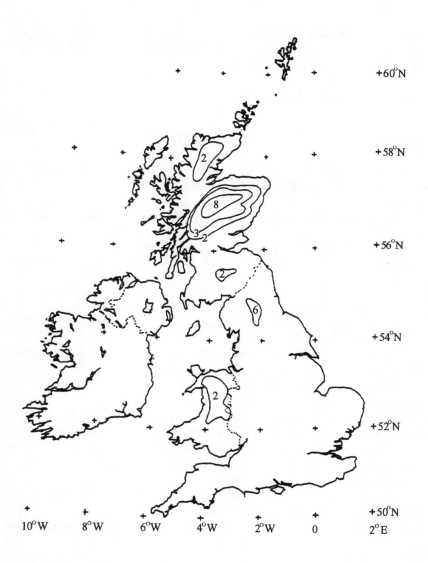

Figure 6.3.6 *Mean number of days with snow lying at 0900 GMT for April*

Example 6.4.1

Estimate the monthly-averaged albedo for Aviemore (57.2°N, 3.8°W) for the month of February.

For a predominantly heather-covered area, $\rho_{snow-free}$ may be taken as 0.14. From Figure. 6.3.3, the average number of days the snow is lying on the ground is 18. Therefore using Eq. (6.4.1) the monthly-averaged albedo is

$$\bar{\rho} = 0.14(10/28) + 0.7\,(18/28) = 0.50$$

References

Barry, R.G. and Chambers, R.E. (1966) A preliminary map of summer albedo over England and Wales. *Q. J. Roy. Met. Soc.* 92, 543.
CEC (1984)
CIBSE (1982) *CIBSE Guide A2*. Chartered Institution of Building Services Engineers, London.
Eagleson, P.S. (1970) *Dynamic Hydrology*. McGraw-Hill, New York.
Greiger, R. (1966) *The Climate Near the Ground*. Harvard University Press, Cambridge, M.A.
Gueymard, C. (1987) An anisotropic solar irradiance model for tilted surfaces and its comparison with selected engineering algorithms. *Solar Energy* 38, 367.
Hay, J.E. (1970) *Aspects of the Heat and Moisture Balance of Canada*. PhD Thesis, University of London.
Hay, J.E. (1976) A revised method for determining the direct and diffuse components of the total short-wave radiation. *Atmosphere* 14, 278.
Hunn, B.D. and Calafell, D.O. (1977) Determination of average ground reflectivity for solar collectors. *Solar Energy* 19, 87.
Iqbal, M. (1983) *Introduction to Solar Radiation*. Academic Press, New York.
Kondratyev, K.Y. (1969) *Radiation in the Atmosphere*. Academic Press, New York.
Kung, E.C., Bryson, R.A. and Lenshow, D.H. (1964) A study of a continental surface albedo on the basis of flight measurements and structure of the Earth's surface cover over North America. *Monthly Rev.* 92, 543.
Liu, B.Y.H. and Jordan, R.C. (1963) The long-term average performance of flat-plate solar-energy collectors. *Solar Energy* 7, 53.
Meteorological Office (1972) *Meteorological Glossary*, ed. D.H. McIntosh. HMSO, London.
Monteith, J.L. (1959) The reflection of short-wave radiation by vegetation. *Q. J. Roy. Met. Soc.* 85, 392.
Monteith, J.L. (1973) *Principles of Environmental Physics*. Edward Arnold, London.
Muneer, T. (1987) *Solar Radiation Modelling for the United Kingdom*. PhD Thesis, Council for National Academic Awards, London.
Oguntoyinbo, J.S. (1970) Reflection coefficient of natural vegetation, crops and urban surfaces in Nigeria. *Q. J. Roy. Met. Soc.* 96, 430.

Saluja, G.S. and Muneer, T. (1988) Estimation of ground-reflected radiation for the United Kingdom. *BSER&T* 9, 189.

Siegel, R. and Howell, J.R. (1972) *Thermal Radiation Heat Transfer*. McGraw-Hill, New York.

Temps, R.C. and Coulson, K.L. (1977) Solar radiation incident upon slopes of different orientations. *Solar Energy* 19, 179.

7 PSYCHROMETRICS

Psychrometrics deals with the study of moist air and its thermodynamic properties. Owing to the wide application of psychrometrics in the design and application of building air-conditioning systems, properties are more commonly known as psychrometric properties. The aim of this chapter is to introduce these properties, present models for their computation and provide the relevant FORTRAN routines.

It may be recalled that in Chapter 3 with reference to the meteorological radiation model it was pointed out that for any given hour, the dew-point temperature was required to determine the prevailing atmospheric moisture content. The commonly recorded psychrometric parameters are either the dry-bulb and wet-bulb temperatures, or the dry-bulb temperature and relative humidity. In the United Kingdom, for example, the former two components are recorded at over 500 locations. In many locations these quantities are reported as the daily maximum and minimum temperatures. However, for the recording sites managed by the Meteorological Office hourly data are easily available.

For those sites where only daily maximum and minimum temperatures are recorded a method is required to enable these to be decomposed into hourly or sub-hourly, values. A procedure and an electronic routine based on ASHRAE (1993) is also provided herein to this effect.

7.1 Psychrometric properties

All models presented in this chapter are those of ASHRAE (1993). Thus the notation of ASHRAE is closely followed in the accompanying text as well as in the electronic routines, presented as Prog7-1.For (given input: dry-bulb and wet-bulb temperatures) and Prog7-2.For (given input: dry-bulb temperature and relative humidity).

If t is the temperature (°C) and T the absolute temperature (K) then $T = t + 273.15$. If W_s is the humidity ratio at saturation, then for a given temperature and pressure, humidity ratio W can have any value from zero to W_s. We define v_a as the specific volume of dry air (m³/kg), h_a the specific enthalpy of dry air (kJ/kg of dry air) and p_{ws} the vapour pressure of water (kPa) in moist air.

The saturation pressure over *ice* for the temperature range -100 to 0°C is given by

$$\ln(p_{ws}) = C_1/T + C_2 + C_3T + C_4T^2 + C_5T^3 + C_6T^4 + C_7\ln(T) \qquad (7.1.1)$$

where $C_1 = -5.674\ 535\ 9 \times 10^3$, $C_2 = -5.152\ 305\ 8 \times 10^{-1}$, $C_3 = -9.677\ 843\ 0 \times 10^{-3}$,

$C_4 = 6.221\ 570\ 1 \times 10^{-7}$, $C_5 = 2.074\ 782\ 5 \times 10^{-9}$, $C_6 = -9.484\ 024\ 0 \times 10^{-13}$ and $C_7 = 4.163\ 501\ 9$.

The range of saturation pressure over *liquid water* for the temperature range of 0 to 200°C is given by

$$\ln(p_{ws}) = C_8/T + C_9 + C_{10}T + C_{11}T^2 + C_{12}T^3 + C_{13}\ln(T) \tag{7.1.2}$$

where $C_8 = -5.800\ 220\ 6 \times 10^3$, $C_9 = -5.516\ 256\ 0$, $C_{10} = -4.864\ 023\ 9 \times 10^{-2}$, $C_{11} = 4.176\ 476\ 8 \times 10^{-5}$, $C_{12} = -1.445\ 209\ 3 \times 10^{-8}$ and $C_{13} = 6.545\ 967\ 3$.

The actual humidity ratio W is given by

$$W = 0.621\ 98\ \frac{p_w}{p - p_w} \tag{7.1.3}$$

where p_w is the partial pressure of water vapour and p is the atmospheric pressure.
Similarly, the humidity ratio at saturation is given by

$$W_s = 0.621\ 98\ \frac{p_{ws}}{p - p_w} \tag{7.1.4}$$

The relative humidity may be obtained from

$$\phi = \left.\frac{p_w}{p_{ws}}\right|_{t,p} \tag{7.1.5}$$

It may easily be shown that

$$v = \frac{R_a T(1 + 1.6078W)}{p} \tag{7.1.6}$$

Where R_a is the gas-constant for water substance.

The enthalpy of a mixture of gases equals the sum of the individual partial enthalpies of the components. Thus

$$h = h_a + W h_g \tag{7.1.7}$$

where h_a is the specific enthalpy for dry air and h_g is the specific enthalpy for saturated water vapour at the temperature of the mixture. ASHRAE (1993) recommends

$$h_a = 1.006\ t \ (\text{kJ/kg}) \tag{7.1.8}$$
$$h_g = 2501 + 1.805\ t \ (\text{kJ/kg}) \tag{7.1.9}$$
$$h = 1.006t + W(2501 + 1.805t) \ (\text{kJ/kg}) \tag{7.1.10}$$

where t is the dry-bulb temperature (°C). For an adiabatic process,

$$h + (W_s^* - W) h_w^* = h_s^* \tag{7.1.11}$$

h_w^* and h_s^* respectively denote the actual and saturation enthalpies of water at the temperature t^*.

The properties of W_s^*, h_w^*, and h_s^* are functions only of the temperature t^* for a fixed value of pressure. The value of t^*, which satisfies Eq. (7.1.11) for given values of h, W and p, is defined as the thermodynamic wet-bulb temperature.

Substituting the approximate relationship $h_w^* = 4.186t^*$ in Eq. (7.1.11) and solving for the humidity ratio yields

$$W = \frac{(2501 - 2.381t^*)W_s^* - (t - t^*)}{2501 + 1.805t - 4.186t^*} \tag{7.1.12}$$

The dew-point temperature t_d is obtainable as follows:

$$t_d = C_{14} + C_{15}\alpha + C_{16}\alpha^2 + C_{17}\alpha^3 + C_{18}(p_w)^{0.1984} \tag{7.1.13}$$

where t_d is in the range 0 to 93 °C. However, when t_d is below 0 °C,

$$t_d = 6.0912\ 60\alpha + 0.4959\alpha^2 \tag{7.1.14}$$

Here $\alpha = \ln p_w$, $C_{14} = 6.54$, $C_{15} = 14.526$, $C_{16} = 0.7389$, $C_{17} = 0.094\ 86$ and $C_{18} = 0.4569$.

ASHRAE (1993) provides detailed, step-by-step computational procedures for all psychrometric properties for a variety of input combinations. The basic equations for the above task are those given above. Use of Prog7-1.For and Prog7-2.For is now demonstrated.

Example 7.1.1

Compute psychrometric properties given dry-bulb and wet-bulb temperatures of 30 and 20 °C.

Prog7-1.For produces the following output:

relative humidity = 0.398 (39.8%)
humidity ratio = 0.0105 kg water vapour per kg dry air
specific volume = 0.873 m³/kg dry air
enthalpy of moist air = 57.1 kJ/kg dry air
dew-point temperature = 14.9 °C

Example 7.1.2

Compute psychrometric properties given a dry-bulb temperature of 30 °C and a relative humidity of 0.398 (39.8%).

Prog7-2.For produces the following output:

humidity ratio = 0.0105 kg water vapour per kg dry air
specific volume = 0.873 m³/kg dry air
enthalpy of moist air = 57.1 kJ/kg dry air
dew-point temperature = 14.9 °C
wet-bulb temperature = 20 °C.

A psychrometric chart prepared with the execution of Prog7-2.For is shown in Figure 7.1.1. The corresponding data are included in File7-1.Csv. The advantage of this type of data base is that it can easily be imported in any of the popular spreadsheet software packages. The user may then superimpose their own computed data (obtained via one of the presently available routines) to produce charts to display their psychrometric state points or processes.

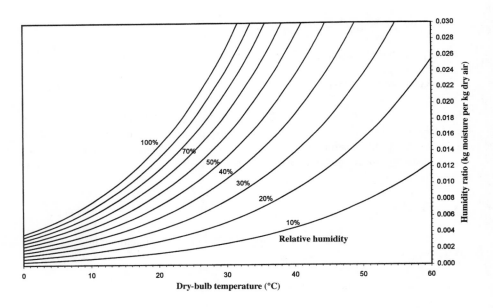

Figure 7.1.1 *Psychrometric chart based on Prog 7-1.For*

7.2 Hourly temperature model

As mentioned above, very often daily or monthly-mean maximum and minimum temperatures are available from weather reports. However, for simulation purposes hourly data are required. Towards this end ASHRAE provides the following simple model:

temperature (at any hour) =
daily maximum temperature − (daily range × percentage) (7.2.1)

The daily range is the difference between the daily maximum and minimum temperatures, and the percentage is the figure corresponding to each hour given in Table 7.2.1.

Table 7.2.1 Diurnal temperature swing

Hour	%	Hour	%
1	87	13	11
2	92	14	3
3	96	15	0
4	99	16	3
5	100	17	10
6	98	18	21
7	93	19	34
8	84	20	47
9	71	21	58
10	56	22	68
11	39	23	76
12	23	24	82

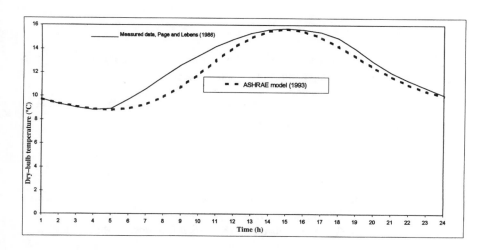

Figure 7.2.1 *Evaluation of ASHRAE hourly temperature model for dry-bulb temperature*

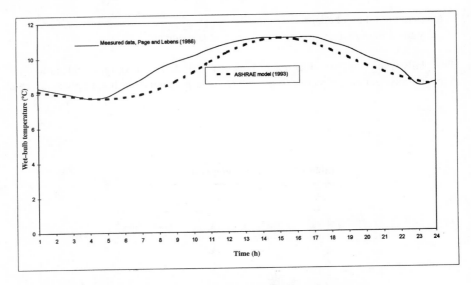

Figure 7.2.2 *Evaluation of ASHRAE hourly temperature model for wet-bulb temperature*

Figures 7.2.1 and 7.2.2 enable evaluation of the performance of the ASHRAE hourly temperature model against measured data for one location in the United Kingdom. The plots show reasonably good agreement, thus making it possible to obtain the hourly values in the absence of such information.

Example 7.2.1

Compute the dry-bulb and wet-bulb temperatures at 21 hours, given the respective daily maximum and minimum temperatures of 15 and 3 °C.

Prog7-3.For enables computation of all 24 hour cyclic temperatures. The temperature for 21 hours thus obtained is 8°C.

References

ASHRAE (1993) *Handbook of Fundamentals*. American Society of Heating, Refrigerating and Air-Conditioning Engineers, Atlanta.

Page, J.K. and Lebens, R. (1986) *Climate in the United Kingdom*. HMSO, London.

PROJECTS

Chapter 1

P1 Daylight illuminance and irradiance data are available as 5 minute averages from world-wide locations given in Appendix A. Merge and edit Prog1-6.For and Prog1-7.For to produce a FORTRAN program which enables computation of daily sunrise and sunset times and the sun's altitude and azimuth at every 5 minute time step, given the latitude and longitude of a location. Enable your program to read/write data via computer files.

P2 Modify Prog1-7.For to produce a year's listing of civil twilight times, given the latitude and longitude of a location.

Chapter 2

P3 Refer to the basic meteorological data available in Appendix B. Using the monthly-averaged sunshine duration, prepare tables of monthly-averaged daily global $\bar{G}$ and diffuse $\bar{D}$ irradiation for the given sites. You may modify Prog2-1.For to enable the input/output flow via data files.

P4 Using values of $\bar{G}$ and $\bar{D}$ obtained in Project P3, prepare tables for monthly-averaged daily vertical irradiation for the four cardinal aspects using Prog2-2.For as a platform.

Chapter 3

P5 Using computations for $\bar{G}$ and $\bar{D}$ obtained in Project P3, prepare tables for monthly-averaged hourly irradiation $\bar{I}_G$ and $\bar{I}_D$. Prog3-1.For may be adopted for this task.

P6 Compute monthly-averaged hourly global and diffuse illuminance using the output obtained from Project P5.

P7 Using Prog3.8.For and closely following the working of Example 3.7.1, prepare monthly frequency distribution tables for horizontal irradiance and illuminance for Manchester, UK for 9, 10, 11 and 12 hours GMT.

Chapter 4

P8 Using the monthly-averaged hourly irradiation, $\bar{I}_G$ and $\bar{I}_D$ obtained in Project P5, compute vertical irradiation for the four cardinal aspects. You may use Prog4-1.For for this task.

P9 Using the data obtained from Project P8, obtain hourly illuminances for the respective slopes. Prog4-2.For may be adapted for these computations.

P10 Compute the luminance distribution of the sky canopy for 9 and 12 hours GMT for London, UK using Prog4-3.For and the corresponding values of $\bar{I}_G$ and $\bar{I}_D$ and obtained in Project P5. How do these distributions compare with the Japanese data given in Section 4.5, Tables 4.5.3–4.5.9.

Chapter 7

P11 Using the ASHRAE hourly temperature model given in Chapter 7, Eq. (7.2.1), compute the hourly dry-bulb and wet-bulb temperatures for London for the period November to February. Use the monthly average maximum/minimum temperature data given in Appendix B.

P12 Plot the above data points on the electronic psychrometric chart which may be drawn using the data given in File7-1.Csv. You may use any of the popular computer spreadsheets for this task.

APPENDIX A: International Daylight Measurement Programme

World-wide measurement stations, classified as follows:

R: research class
sR: simplified research class
G: general class
S: simplified general class

Code	Country	Location	Class
J1	Japan	Fukuoka	R
J2	Japan	Kyoto	R
J3	Japan	Sapporo	R
J4	Japan	Uozo	G
J5	Japan	Nagoya (Daido)	G
J6	Japan	Nagoya (Meijo)	G
J7	Japan	Toyota	G
J8	Japan	Suita	S
J9	Japan	Ashikaga	sR
J10	Japan	Tokyo	R
J11	Japan	Chofu	R
J12	Japan	Tshukuba	R
J13	Japan	Kiyose	S
J14	Japan	Osaka	S
GB1	UK	Garston	R
GB2	UK	Edinburgh	G
GB3	UK	Manchester	R
GB4	UK	Sheffield	G
F1	France	Nantes	G
F2	France	Vauix en Velin	G
F3	France	Strasbourg	S
F4	France	Chanbery	S
F5	France	Grenoble	S
S1	Sweden	Norrköping	R
S2	Sweden	Gävle	G
S3	Sweden	Kiruna	G
RFA1	Germany	Hamburg	G
RFA2	Germany	Freiburg	S
CH1	Switzerland	Geneva	R
NL1	Netherlands	Eindhoven	R
G1	Greece	Athens	G
P1	Portugal	Lisbon	G
IL1	Israel	Bet Dagan	S
SU1	Russia	Voelkovo	G
SU2	Russia	Moscow	S
SU3	Ukraine	Karadag	S
USA1	USA	Ann Arbor	R
USA2	USA	Albany	G
USA3	USA	Cape Canaveral	R
CDN1	Canada	Calgary	G
AUS1	Australia	Sydney	R
RC1	China	Chongquing	G
RC2	China	Beijing	R
RC3	China	Changchun	G
SGP1	Singapore	Singapore	R
SGP2	Singapore	Singapore	G
IND1	India	Roorkee	R

Source: H. Nakamura, Chair CIE-TC3.25.

APPENDIX B: Mean–monthly weather data for selected UK sites

Location	Eq. (2.4.1) a	a'	Lat. N	Long. W	SH	T max.	T min.	T_{wb} max.	T_{wb} min.	Eq. (2.4.1) b	SH	T max.	T min.	T_{wb} max.	T_{wb} min.	Eq. (2.4.1) b
					January						*February*					
Aberdeen	0.24	0.15	57.150	2.100	1.61	4.7	-0.2	3.2	-0.2	0.51	2.82	5.4	-0.2	3.3	-0.2	0.51
Dundee	0.24	0.14	56.483	3.000	1.66	5.1	0.1	3.3	0.1	0.51	2.87	5.9	0.2	3.5	0.2	0.51
Edinburgh	0.23	0.14	55.950	3.200	1.33	5.9	0.0	3.4	0.0	0.51	2.42	6.6	0.2	3.4	0.2	0.51
Glasgow	0.23	0.14	55.867	4.233	1.15	5.5	0.8	3.6	0.8	0.52	2.20	6.3	0.8	3.7	0.8	0.52
Newcastle	0.24	0.15	54.983	1.583	1.37	5.6	2.2	3.2	2.2	0.49	2.36	6.0	2.1	3.5	2.1	0.49
Belfast	0.25	0.15	54.583	5.917	1.43	5.8	1.3	4.0	1.3	0.49	2.40	6.5	1.1	4.1	1.1	0.49
Manchester	0.25	0.16	53.500	2.250	1.27	5.5	1.2	3.8	1.2	0.48	2.19	6.3	1.0	4.0	1.0	0.48
Liverpool	0.25	0.16	53.417	2.917	1.37	6.1	2.1	4.0	2.1	0.48	2.24	6.6	1.9	4.0	1.9	0.48
Sheffield	0.25	0.16	53.383	1.500	1.23	5.6	1.2	3.2	1.2	0.48	1.93	6.0	1.1	3.5	1.1	0.48
Leicester	0.25	0.17	52.633	1.083	1.35	5.8	0.1	3.5	0.1	0.47	2.25	6.5	0.3	3.6	0.3	0.47
Birmingham	0.25	0.17	52.500	1.833	1.38	5.1	1.5	3.5	1.5	0.47	2.03	5.7	1.3	3.6	1.3	0.47
London	0.23	0.14	51.500	0.167	1.54	6.1	2.3	4.7	2.3	0.45	2.28	6.8	2.3	5.0	2.3	0.45
Bristol	0.26	0.17	51.430	2.580	1.45	6.3	1.6	4.3	1.6	0.47	2.24	6.9	1.5	4.4	1.5	0.47
Cardiff	0.26	0.17	51.483	3.217	1.60	6.8	1.5	4.5	1.5	0.48	2.64	6.9	1.5	4.5	1.5	0.48
Plymouth	0.25	0.16	50.383	4.167	1.87	8.0	3.8	6.2	3.8	0.50	2.86	7.9	3.2	6.1	3.2	0.50
					March						*April*					
Aberdeen	0.24	0.15	57.150	2.100	3.43	7.7	1.2	4.4	1.2	0.51	5.09	1.1	0.3	6.2	0.3	0.60
Dundee	0.24	0.14	56.483	3.000	3.29	8.2	1.8	5.1	1.8	0.51	5.02	11.5	3.5	6.8	3.5	0.59
Edinburgh	0.23	0.14	55.950	3.200	3.2	8.7	1.8	5.0	1.8	0.51	4.98	11.7	3.0	6.9	3.4	0.59
Glasgow	0.23	0.14	55.867	4.233	3.04	8.8	2.2	5.1	2.2	0.52	4.90	1.2	0.4	7.3	0.4	0.58
Newcastle	0.24	0.15	54.983	1.583	3.29	7.8	3.1	5.2	3.1	0.49	5.01	1.0	0.5	7.4	0.5	0.58
Belfast	0.25	0.15	54.583	5.917	3.30	9.0	2.4	5.4	2.4	0.49	5.12	1.2	0.4	7.5	0.4	0.56
Manchester	0.25	0.16	53.500	2.250	3.52	9.2	2.3	5.7	2.3	0.48	4.79	1.2	0.4	7.9	0.4	0.56
Liverpool	0.25	0.16	53.417	2.917	3.53	9.5	3.1	5.8	3.1	0.48	4.87	12.4	5.3	8.1	5.3	0.55
Sheffield	0.25	0.16	53.383	1.500	2.98	8.7	2.3	5.2	2.3	0.48	4.3	1.2	0.5	7.4	0.5	0.56
Leicester	0.25	0.17	52.633	1.083	3.24	9.4	1.5	5.2	1.5	0.47	4.67	12.9	3.8	7.6	3.8	0.54
Birmingham	0.25	0.17	52.500	1.833	3.14	8.7	2.6	5.2	2.6	0.47	5.55	1.2	1.5	7.6	1.5	0.54
London	0.23	0.14	51.500	0.167	3.62	9.8	3.4	6.5	3.4	0.45	5.41	1.3	0.6	8.9	0.6	0.52
Bristol	0.26	0.17	51.430	2.580	3.55	9.9	2.6	6.2	2.6	0.47	5.15	13.1	4.8	8.6	4.8	0.54
Cardiff	0.26	0.17	51.483	3.217	3.95	10.0	2.8	5.9	2.8	0.48	5.44	1.3	0.5	8.0	0.5	0.54
Plymouth	0.25	0.16	50.383	4.167	4.23	9.9	4.3	7.1	4.3	0.50	6.07	1.2	0.6	8.8	0.6	0.54
					May						*June*					
Aberdeen	0.24	0.15	57.150	2.100	5.51	12.7	5.0	9.0	5.0	0.60	5.80	16.1	7.9	12.0	7.9	0.60
Dundee	0.24	0.14	56.483	3.000	5.44	14.3	5.8	9.2	5.8	0.59	5.60	17.6	8.9	12.4	8.9	0.59
Edinburgh	0.23	0.14	55.950	3.200	5.69	14.2	6.0	9.4	6.0	0.59	6.08	17.3	9.1	12.5	9.1	0.59
Glasgow	0.23	0.14	55.867	4.233	5.95	15.1	6.2	10.3	6.2	0.58	6.04	17.9	9.3	13.2	9.3	0.58
Newcastle	0.24	0.15	54.983	1.583	5.50	12.2	7.1	10.5	7.1	0.58	6.01	15.7	10.1	13.8	10.1	0.58
Belfast	0.24	0.15	54.583	5.917	6.17	14.1	6.1	10.3	6.1	0.56	5.82	17.4	9.2	13.2	9.2	0.56
Manchester	0.25	0.16	53.500	2.250	5.93	15.7	6.9	11.3	6.9	0.56	6.37	18.8	9.9	14.3	9.9	0.56
Liverpool	0.25	0.16	53.417	2.917	5.91	15.7	7.8	10.8	7.8	0.55	6.57	18.7	10.9	13.8	10.9	0.55
Sheffield	0.25	0.16	53.383	1.500	5.24	15.4	7.2	10.5	7.2	0.56	6.11	18.7	10.3	13.8	10.3	0.56
Leicester	0.25	0.17	52.633	1.083	5.75	16.1	6.1	10.8	6.1	0.54	6.22	19.3	9.1	14.1	9.1	0.54
Birmingham	0.25	0.17	52.500	1.833	5.47	15.4	7.4	10.8	7.4	0.54	6.15	18.7	10.4	14.1	10.4	0.54
London	0.23	0.14	51.500	0.167	6.56	16.8	8.4	12.3	6.4	0.52	7.14	20.2	11.5	15.7	11.5	0.52
Bristol	0.26	0.17	51.430	2.580	6.14	16.1	7.3	11.3	7.3	0.54	6.8	19.2	10.6	14.4	10.6	0.54
Cardiff	0.26	0.17	51.483	3.217	6.34	16.2	7.7	11.0	7.7	0.54	6.92	19.2	10.7	14.0	10.7	0.54
Plymouth	0.25	0.16	50.383	4.167	7.04	14.9	8.2	11.4	8.2	0.54	7.27	17.6	11.1	14.2	11.1	0.54

SH mean daily sunshine hours
T, max./min. mean daily maximum/minimum temperatures
T_{wb}, max./min. mean daily maximum/minimum wet-bulb temperatures

Location	Eq. (2.4.1) a	a'	Lat. N	Long. W	SH	T max.	T min.	T_{wb} max.	T_{wb} min.	Eq. (2.4.1) b	SH	T max.	T min.	T_{wb} max.	T_{wb} min.	Eq. (2.4.1) b
					July						August					
Aberdeen	0.24	0.15	57.150	2.100	1.61	17.5	9.6	13.7	9.6	0.60	2.82	17.1	9.4	13.6	9.4	0.60
Dundee	0.24	0.14	56.483	3.000	5.08	19.0	10.6	13.1	9.9	0.59	4.44	18.3	10.3	13.4	10.2	0.59
Edinburgh	0.23	0.14	55.950	3.200	5.43	18.6	10.8	13.0	9.9	0.59	4.75	18.2	10.6	13.4	10.3	0.59
Glasgow	0.23	0.14	55.867	4.233	5.14	18.6	10.8	14.3	10.8	0.58	4.60	18.5	10.6	14.2	10.6	0.58
Newcastle	0.24	0.15	54.983	1.583	5.31	17.4	12.0	15.2	12.0	0.58	4.68	17.2	12.0	15.0	12.0	0.58
Belfast	0.25	0.15	54.583	5.917	4.41	18.1	10.7	14.3	10.7	0.56	4.42	18.0	10.5	14.2	10.5	0.56
Manchester	0.25	0.16	53.500	2.250	5.00	19.6	11.7	15.4	11.7	0.56	4.90	19.4	11.5	15.4	11.5	0.56
Liverpool	0.25	0.16	53.417	2.917	5.28	19.6	12.8	15.1	12.3	0.55	4.94	19.5	12.5	14.6	12.2	0.55
Sheffield	0.25	0.16	53.383	1.500	5.15	19.8	12.2	15.2	12.2	0.56	4.58	19.5	11.9	15.0	11.9	0.56
Leicester	0.25	0.17	52.633	1.083	5.43	20.7	11.2	15.7	11.2	0.54	4.96	20.4	10.7	15.3	10.7	0.54
Birmingham	0.25	0.17	52.500	1.833	5.29	19.9	12.1	15.7	12.1	0.54	4.86	19.5	11.9	15.3	11.9	0.54
London	0.23	0.14	51.500	0.167	6.34	21.6	13.4	17.2	13.4	0.52	5.90	21.0	13.1	16.7	13.1	0.52
Bristol	0.26	0.17	51.430	2.580	6.07	20.6	12.4	15.1	12.3	0.54	5.5	20.1	12.1	15.0	0.0	0.58
Cardiff	0.26	0.17	51.483	3.217	6.11	20.4	12.3	15.5	12.3	0.54	5.77	20.1	12.3	15.5	12.3	0.54
Plymouth	0.25	0.16	50.383	4.167	6.53	19.0	12.7	15.7	12.7	0.54	6.09	19.0	12.7	15.8	12.7	0.54
					September						October					
Aberdeen	0.24	0.15	57.150	2.100	3.43	15.5	8.0	11.3	8.0	0.60	5.09	12.4	5.8	9.1	5.8	0.51
Dundee	0.24	0.14	56.483	3.000	3.92	16.4	8.7	12.3	8.7	0.59	2.93	12.9	6.1	10.0	6.1	0.51
Edinburgh	0.23	0.14	55.950	3.200	4.00	16.6	9.0	12.3	9.0	0.59	2.94	13.3	6.5	9.9	6.5	0.51
Glasgow	0.23	0.14	55.867	4.233	3.53	16.3	9.1	12.3	9.1	0.58	2.45	13.0	6.8	9.8	6.8	0.52
Newcastle	0.24	0.15	54.983	1.583	4.03	15.9	10.6	13.0	10.6	0.58	3.00	13.0	8.4	10.0	8.4	0.49
Belfast	0.25	0.15	54.583	5.917	3.47	16.0	9.3	12.4	9.3	0.56	2.53	12.9	7.2	10.2	7.2	0.49
Manchester	0.25	0.16	53.500	2.250	3.97	17.3	10.0	13.4	10.0	0.56	2.98	13.7	7.3	10.5	7.3	0.48
Liverpool	0.25	0.16	53.417	2.917	3.89	17.5	11.0	13.5	11.0	0.55	2.79	14.0	8.4	11.1	8.4	0.48
Sheffield	0.25	0.16	53.383	1.500	3.71	17.4	10.3	13.0	10.3	0.56	2.71	13.7	7.6	10.0	7.6	0.48
Leicester	0.25	0.17	52.633	1.083	3.99	18.0	9.1	13.1	9.1	0.54	2.92	14.2	6.3	10.1	6.3	0.47
Birmingham	0.25	0.17	52.500	1.833	3.92	17.5	10.4	13.1	10.4	0.54	2.81	13.5	7.8	10.1	7.8	0.47
London	0.23	0.14	51.500	0.167	4.77	18.5	11.4	14.5	11.4	0.52	3.29	14.7	8.5	11.6	8.5	0.45
Bristol	0.26	0.17	51.430	2.580	4.29	18.0	10.3	13.9	10.3	0.54	3.01	14.5	7.7	11.6	7.7	0.47
Cardiff	0.26	0.17	51.483	3.217	4.43	18.0	10.7	13.8	10.7	0.54	3.28	14.5	8.0	11.0	8.0	0.48
Plymouth	0.25	0.16	50.383	4.167	4.93	17.5	11.6	14.3	11.6	0.54	3.63	14.8	9.4	12.1	9.4	0.50
					November						December					
Aberdeen	0.24	0.15	57.150	2.100	5.51	8.0	2.6	5.2	2.6	0.51	5.80	5.7	1.0	4.4	1.0	0.51
Dundee	0.24	0.14	56.483	3.000	2.04	8.4	2.6	5.4	2.6	0.51	1.47	6.2	1.0	3.1	1.0	0.51
Edinburgh	0.23	0.14	55.950	3.200	1.72	9.0	2.6	5.4	2.6	0.51	1.19	7.0	1.1	3.3	1.1	0.51
Glasgow	0.23	0.14	55.867	4.233	1.55	8.7	3.3	5.5	3.3	0.52	0.98	6.5	1.9	4.3	1.9	0.52
Newcastle	0.24	0.15	54.983	1.583	1.75	8.8	5.1	6.0	5.1	0.49	1.24	6.6	3.2	3.9	3.0	0.49
Belfast	0.25	0.15	54.583	5.917	1.87	8.9	3.9	6.2	3.9	0.49	1.12	6.7	2.5	4.8	2.5	0.49
Manchester	0.25	0.16	53.500	2.250	1.64	8.9	4.0	6.4	4.0	0.48	1.21	6.4	2.2	4.4	2.2	0.48
Liverpool	0.25	0.16	53.417	2.917	1.68	9.5	5.1	6.6	5.1	0.48	1.3	7.1	3.1	4.1	3.1	0.48
Sheffield	0.25	0.16	53.383	1.500	1.45	8.9	4.3	6.0	4.3	0.48	1.14	6.9	2.3	3.9	2.3	0.48
Leicester	0.25	0.17	52.633	1.083	1.7	9.3	3.1	6.2	3.1	0.47	1.22	7.0	1.1	4.0	1.1	0.47
Birmingham	0.25	0.17	52.500	1.833	1.61	8.6	4.5	6.2	4.5	0.47	1.35	6.2	2.7	4.0	2.7	0.47
London	0.23	0.14	51.500	0.167	1.92	9.5	5.3	7.5	5.3	0.45	1.38	7.2	3.4	5.5	3.4	0.45
Bristol	0.26	0.17	51.430	2.580	1.83	9.7	4.5	7.1	4.5	0.47	1.44	7.3	2.9	4.4	2.9	0.47
Cardiff	0.26	0.17	51.483	3.217	2.01	10.2	4.7	7.4	4.7	0.48	1.55	8.0	2.7	5.4	2.7	0.48
Plymouth	0.25	0.16	50.383	4.167	2.30	11.1	6.4	8.8	6.4	0.50	1.82	9.2	4.9	7.0	4.9	0.50

SH mean daily sunshine hours
T, max./min. mean daily maximum/minimum temperatures
T_{wb}, max./min. mean daily maximum/minimum wet-bulb temperatures

APPENDIX C:
Instructions for use of Fortran programs

All but five FORTRAN programs included with this text are free-standing modules. The input for these programs is to be provided via keyboard. They have been designed with simplicity in mind. The programs do not perform quality control checks on the user's input. As such it is important that all the requisite data are provided with care.

Many of these programs can be used as building blocks for the user's specific needs. Six of the accompanying FORTRAN programs require input files which are included in the CD in the Csv format (the advantage of this format is that the files may be directly imported in Excel or Lotus 1-2-3 spreadsheets). These programs with their associated input/output files are listed below.

Four more data files, i.e. File2-1a.Csv, File2-1b.Csv, File3-1.Csv and File7-1.Csv are provided for general purpose calculations. The titles of these data files are given at the beginning of the book.

FORTRAN differentiates between real and integer numbers and therefore care must be taken when keying in the required data. Clear instructions are provided at each stage of execution of the respective programs.

Program	*Input file*	*Output file*
Prog3-9.For	File1-1.Csv	Out3-9.Dat
Prog3-2.For	In3-2.Csv	Out3-2.Dat
Prog3-5.For	In3-5.Csv	—
Prog4-3.For	In4-3.Csv	Out4-3.Dat
Prog4-4.For	In4-3.Csv	Out4-4.Dat
Prog5-1.For	File5-1.Csv	Out5-1.Dat

INDEX

Aerosols:
    scattering by, and Ångström turbidity formula, 63-4, 157
    solar radiation through, 62
Air-conditioning efficiency xxiii-xxiv
Albedo, see Ground albedo
All-sky distributions, radiance and luminance distributions on sloping surfaces, 142-5
Ångström regression equation, 28
Ångström turbidity formula/coefficient, 63-4, 77, 78, 157
Annual irradiation data for world-wide locations, 36
Annual-averaged diffuse irradiation, 34-7
Apparent solar time (AST), 2, 5
ASHREA models, 183-4, 184-6
AST, see Apparent solar time
Atmospheric transmittances:
    according to Bird and Hulstrom, 64
    according to Davies et al., 64
    according to Iqbal, 64
    according to Lacis and Hansen, 64, 65
Average luminous efficacy models, 82-3

Beam irradiation on a slope, 32
Beam luminous efficacy models, 78
Beam solar radiation, attenuation of, 63
Bougher's law for flux arriving at earth's surface, 155-6
BRE sky component tables, 89
British Summer Time (BST), estimating value of, 99-100
Building materials, albedo of, 167

Campbell-Stokes sunshine recorder, 18
Canada, ground albedo, 167
CIE, see Commission Internationale de l'Éclairage
Circumsolar sky-diffuse irradiance model, 111-12
Clean Air Acts, xxiii
Clear sky:
    luminance distribution, 138
    radiance distribution, 138
Clearness function, 110-11
Clouds, effect of, 77-8
CO_2 production, xxvii
Coefficient of correlation, 22, 23
Commission Internationale de l'Éclairage (CIE):
    International Daylight Measurement Programme and Year (1991), xxi, 1, 18, 189
    and standardized sensitivity of daylight, 17

Daily fractional sunshine, UK, 30
Daily horizontal diffuse irradiation:
    annual-averaged, 34-7
    correlation studies:
        in Canada, 38-9
        in India, 38
        in Italy, 39
        in UK, 39
        in USA 38, 41
    daily and monthly inequalities, 42-5
    desert and tropical location,s 33
    Indian regression curve, 33
    monthly-averaged, 32-4
    regression equations/curves, 38-42
    slope irradiation, 32
Daily horizontal global irradiation:
    correlation with hourly global irradiation, 53-5
    Cowley's model/equation, 37-8
    monthly-averaged, 28-32, 34
    on a slope, 32
Daily irradiation:
    introduction, 27-6
    see also Annual-averaged diffuse irradiation; Daily horizontal diffuse irradiation; Daily horizontal global irradiation; Monthly-averaged hourly horizontal diffuse irradiation; Monthly-averaged hourly horizontal global irradiation
Daily slope irradiation, 32, 45-9
    for Easthamstead, UK, 47
    from Liu and Jordan, 45-6

194 INDEX

from Muneer and Saluja, 46-7
for Lerwick, UK, 47, 49
Day number (DN), 2
  FORTRAN routine for, 2
Daylight availability, 98
Daylight factor, 89-91
Daylight illuminance distributions, 96-7
Daylight luminous efficacy, 79
Daylight Measurement Programme, International, 189
Daylight measurement/sensors, 16, 17
  see also Illuminance; Pyranometers
DEC, see Solar declination angle
Diffuse irradiance, measurement of, 17-18
Diffuse irradiation, see Daily horizontal diffuse irradiation
Diffuse luminous efficacy models, 78-80
Diffuse radiation, 63, 111
Diffuse to beam ratio (DBR), 68
Distance between locations, 15-16
Diurnal cycle, 2
Diurnal duration of bright sunshine, 18
Diurnal temperature swing, 185
DuMortier–Perraudeau–Page luminous efficacy model, 83-4

Electrical consumption saving, 76
Energy availability modelling, xxiv
EOT, see Equation of time
Equation Naturales, 4
Equation of time (EOT),
  accuracy evaluation, 4-5
  FORTRAN 77 routines for, 4
  from Lamm, 3-4
  from Woolf, 3
  tables for year 2002 with solar declination angle, 6-7
European Court of Human Rights building, 76
Eye sensitivity, 15

FORTRAN:
  FORTRAN environment, 1
  FORTRAN-77, 1
  FORTRAN-90, 1
  instructions for use of programs, 192
  use of in this book, xxv
Foster sunshine recorder, 18
Frequency distribution of illuminance, 96-100
Frequency distribution of irradiation, 92-6

Gases, mixed, absorption of radiation by, 159

Glazing, luminance transmission through, 147-8
  multiple glazed windows, 148
Global irradiation, 33
  see also Daily horizontal global irradiation
Global luminous efficacy of daylight, 76
Global luminous efficacy models, 78-80
Global warming, xxii-xxiii
Goudriaan's analysis, 137
Ground albedo (ground-reflected radiation):
  building materials, 167
  in Canada, 167
  explanation of, 165
  grass, 169-10
  ground coverings, 166, 167
  models for, 169-10
  monthly-averaged estimates, 171
  natural surfaces, 166
  snow, 168
  soil covers, 166
  UK maps/atlas, 170-7
  in USA, 168, 169
  vegetative covers, 166
Ground-reflected radiation, see Ground albedo
Gueymard's formulation absorption by water calculations, 160
Gueymard's sky-diffuse irradiance model, 118-20, 125
  overcast sky condition, 119
  part-overcast sky condition, 119-20

Hay's sky-diffuse irradiance model, 113, 125
History of climatical study of radiation, xxi
Horizontal global irradiation, see Daily horizontal global irradiation; Hourly horizontal global irradiation
Hourly horizontal diffuse irradiation:
  correlation with global irradiation, 57-9
  monthly-averaged, for Indian locations, 57
  monthly-averaged, for Eskdalemuir, 56
  MRM model, 75-6
  regression models for:
    by Boes, 74
    by Bugler, 74
    by Erbs, 71
    by Jeter and Balaras, 74
    by Muneer et al., 73
    by Orgill and Hollands, 71
    by Perez et al., 74
    by Reindl, 75
    by Spencer, 73
    Maxwell model, 74-5
Hourly horizontal global irradiation:
  correlation with diffuse irradiation, 57-9
  meteorological radiation model (MRM), calculation with, 59-71

monthly-averaged, correlation with daily global irradiation, 53-5
Hourly horizontal illuminance:
average luminous efficacy models, 82-3
beam luminous efficacy models, 78
clouds, 77-8
Delaunay and Muneer findings, 84-5
Drummond and Ångström equation, 78-9
DuMortier–Perraudeau–Page model, 83-4
global and diffuse luminous efficacy models, 78-80
global luminous efficacy of daylight, 76-7
Japanese locations, 85-6
Linke turbidity factor, 77
Littlefair model, 80-1
Moon's empirical spectral distribution curves, 79
Muneer and Angus evaluations, 81
Navvab et al., illuminance turbidity coefficient, 77, 78
Perez et al. model, 81
recent developments, 76
single averaged value of luminous efficacy model, 78
soot and dust effects, 77
UK Building Regulations, 76
water vapour effects, 77
world-wide variation, 86
zenith luminance, 86-8
Hourly slope beam irradiance, 110
Hourly slope irradiation and illuminance, 109-54
Hourly-averaged beam irradiance, 63

Illuminance:
frequency distribution of, 96-100
measurement of, 17-18
slope beam, 110
Illumination, definition, xxi
INC (sun's angle of incidence), 10, 11-12
India:
clearness index, 29-30
frequency distribution of irradiation, 92-3
monthly-averaged diffuse ratio against clearness index, 32-3
Instantaneous beam irradiance, 63, 110
Intermediate sky distributions, radiance and luminance distributions on sloping surfaces, 138-42
International Daylight Measurement Programme (IDMP) and Year (1991) xxi, 1, 18, 189
Irradiance:
definition, xxi-xxii
slope beam, 110
Irradiation:

definition, xxv
frequency distribution of, 92-6
see also Annual-averaged diffuse irradiation; Daily horizontal diffuse irradiation; Daily horizontal global irradiation; Monthly-averaged hourly horizontal diffuse irradiation
Isotropic sky-diffuse irradiance model, 111, 123-4

Julian day number (JDN), 2-3

Kew recorded daylight data, xxiii
Kipp solarimeters, xxi
Klucher's sky-diffuse irradiance model, 112-13

Lambert's law for energy flux arriving at earth's surface, 155-6
Lighting efficiency, 76
Linke turbidity factor, 77
Littlefair global luminous efficacy model, 80-1, 86
Local civil time (LCT), 2
LONG (longitude of a given locality), 5, 10
Longitude of the standard time meridian (LSM), 5
LSM, *see* Longitude of the standard time meridian
Luminance distributions on sloping surfaces, *see* Radiance and luminance distributions on sloping surfaces
Luminous efficacy models/estimates
beams, 78
by DuMortier–Perraudeau–Page, 83-4
by Littlefair, 80-1
by Perez et al., 81-3
global and diffuse, 78-80
miscellaneous, 84-6

Mardaljevic's comparisons, radiance and luminance distributions on sloping surfaces, 136
Mean bias error (MBE), 23
and the Meteorological radiation model (MRM), 69
Mean-monthly weather data for selected UK sites, 190
Meteorological Office UK:
daily fractional sunshine records, 30
network, 27
Meteorological radiation model (MRM):
accuracy evaluation, 67
at an hourly level, 66

INDEX

for clear skies, 66, 67
for daily- and monthly-averaged irradiation, 70-1
hourly horizontal global irradiation calculations, 59-71
for non-overcast skies, 68-70
for overcast skies, 66-8
overview, 65
Mie scattering, 63, 156, 157
Moist air, *see* Psychrometrics
Monochromatic extinction coefficient, 156
Monochromatic solar spectral radiation, 155-6
Monthly-averaged daily horizontal diffuse irradiation, 32-4
see also Daily horizontal diffuse irradiation
Monthly-averaged daily horizontal global irradiation, 28-32, 34
Monthly-averaged horizontal daily extraterrestrial irradiation, 29
Monthly-averaged hourly horizontal diffuse irradiation, 55-9
Monthly-averaged hourly horizontal global irradiation, 53-5
Moon, illuminance from, 14
Moon and Spencer's sky-diffuse irradiance model, 114
MRM, *see* Meteorological radiation model
Muneer and Angus's slope illuminance model, 134
Muneer's sky-diffuse irradiance model, 121-3, 126-7

National Physics Laboratory in Teddington recorded data, xxiii
Natural surfaces, albedo of, 166
Near infrared (NIR) region, 59-60

Optical air mass, relative and absolute, 62-3
Overcast sky, radiance distribution, 136-8
Ozone:
  absorption of radiation by, 158
  solar radiation through, 60-1

Perez et al.'s sky-diffuse irradiance model, 120-1, 126
Perez et al.'s slope illuminance model, 134
Perez luminous efficacy model, 81-3
Photometers, 17
  see also Illuminance
Photosynthesis, xxii
Projects (for chapters 1, 2, 3, 4, and 7), 187-8
Psychrometrics:
  application of, 181

ASHREA models, 183-6
hourly temperature model, 184-6
psychrometric properties, 181-4
Pyranometers, 16-20
  azimuth error, 19
  CM-11a 16-17, 19-20
  cosine error, 19
  error and uncertainty, 18-20
  Robitzch actinographs, 20

Radiance and luminance distributions on sloping surfaces, 136-48
  all-sky distributions, 142-5
  clear sky distributions, 138
  computer usage for 136, 143-5
  intermediate sky distributions, 138-42
  Mardaljevic's comparisons, 136
  overcast sky distributions, 136-8
  Perez all-sky model, 143, 144
Radiation, eye comparison measurements, 17
Radiation models, *see* Solar radiation models
Rayleigh scattering, 63, 78, 84, 156
Reflected radiation, 111
Reindl et al.'s sky-diffuse irradiance model, 114, 125
Robitzch actinographs, 20
Root mean square error (RMSE), 23

Skartveit and Olseth's sky-diffuse irradiance model, 113-14, 125
Sky clarity indices, 110-11
Sky component, 89
Sky-diffuse irradiance models,
  circumsolar model, 111-12
  Gueymard's model, 118-20, 125
  Hay's model, 113, 125
  isotropic model, 111, 123-4
  Klucher's model, 112-13
  Moon and Spencer's model, 114
  Muneer's model, 121-3, 126-7
  Perez et al.'s model, 120-1, 126
  Reindl et al.'s model, 114, 125
  Skartveit and Olseth's model, 113-14, 125
  Steven and Unsworth's model, 114-18
  Temps and Coulson's clear sky model, 112
Sky-diffuse irradiance on a tilted surface, equation derivation, 115-18
Slope beam irradiance and illuminance, 110
Slope illuminance models:
  CIE overcast sky daylight factor concept, 133
  Muneer and Angus's model, 134
  Perez et al.'s model, 134
Slope irradiation, *see* Daily slope irradiation
Snow, ground albedo, 168, 170

Soil covers, albedo of 166
SOLALT (sun's elevation above horizon), 10
   and hourly horizontal diffuse irradiation, 75
   and sunrise/sunset, 13
Solar altitude, 97
Solar constant, 29
Solar day, 2-3
Solar declination angle (DEC), 8-10
   from Boes, 8
   from Duffett-Smith, 8
   from FORTRAN 77 routines, 9
   from Yallop's algorithm, 8-9
   and monthly–averaged daily horizontal global irradiation, 29
   table for 21st day of each month, 10
Solar radiation measurement, 16-20
   see also Pyranometers
Solar radiation models:
   coefficient of correlation r, 22, 23
   coefficient of determination r^2, 22
   mean bias error (MBE), 23
   regression models and correlation equations, 21
   root mean square error (RMSE), 23
   software packages, 21
   statistical evaluation of, 20-3
   Student's t-distribution, 22
Solar radiation transmission:
   and relative and absolute optical air mass, 62-3
   through aerosols, 62
   through earth's atmosphere, 60-3
   through mixed gases, 60
   through ozone, 60-1
   through water vapour, 61-2
Solar radiations, concepts of, xxi-xxii
Solar spectral radiation:
   absorption by mixed gases, 159
   absorption by ozone, 158
   absorption by water vapour, 159-60
   aerosol (Mie) scattering, 157
   monochromatic, 155-6
   Rayleigh scattering, 156
   spectral radiation model (SRM), 161-2
Solar spectrum, and the MRM, 59-60
Solar sunrise/sunset, 29
Solar time, 5, 8
SOLAZM (Sun's azimuth from north), 10
   and sunrise/sunset 13
Spectral radiation model (SRM), 161-2
Steven and Unsworth's sky-diffuse irradiance model, 114-18
Student's t-distribution, 22, 23

Sunrise:
   actual, 13
   astronomical, 13
Sunset:
   actual, 13
   astronomical, 13
Sunshine fraction (SF), 68
Sunshine recorders:
   Campbell-Stokes, 18
   Foster sunshine switch, 18

Temps and Coulson's clear sky sky-diffuse irradiance model, 112
Tilted surface sky-diffuse irradiance, equation derivation, 115-18
TLT (tilt of a sloped surface), 10, 11-12
Twilight, 13-15
   civil and nautical 14

UK:
   albedo atlas for, 170-7
   frequency distribution of irradiation, 93-4
   mean-monthly weather data for selected sites, 190-1
UK Building Regulations, 76
Ultraviolet (UV) radiation, 59-60
Universal Time (UT), 3
US Standard Atmosphere, 64, 156
USA:
   frequency distribution of irradiation, 93-4
   ground albedo for 168, 169

Vegetative covers, albedo of, 166

Wall azimuth angle (WAZ), 10
Water vapour, *see* Psychrometrics
Water vapour:
   absorption of radiation by, 159-60
   solar radiation through, 61-2
Windows *see* Glazing, luminance transmission through
World Meteorological Organisation (WMO), 19

Yallop's algorithm for EOT, 4

Zenith luminance, 86-8

Read Ch 4, 4.5 4.54

Ex 4.5.1
4.6)
4.b2
4.b3

Ex 4.7